APPROXIMATION AND REGULARISATION METHODS FOR OPERATOR-FUNCTIONAL EQUATIONS

Series on Advances in Mathematics for Applied Sciences – Vol. 95

APPROXIMATION AND REGULARISATION METHODS FOR OPERATOR-FUNCTIONAL EQUATIONS

Denis Sidorov
Russian Academy of Sciences, Russia

Edixon Rojas
Universidad Nacional de Colombia, Colombia

Alexander Sinitsyn
Universidad Nacional de Colombia, Colombia

Nikolai Sidorov
Irkutsk State University, Russia

World Scientific

NEW JERSEY · LONDON · SINGAPORE · BEIJING · SHANGHAI · TAIPEI · CHENNAI

Published by

World Scientific Publishing Co. Pte. Ltd.

5 Toh Tuck Link, Singapore 596224

USA office: 27 Warren Street, Suite 401-402, Hackensack, NJ 07601

UK office: 57 Shelton Street, Covent Garden, London WC2H 9HE

Library of Congress Control Number: 2024041761

British Library Cataloguing-in-Publication Data
A catalogue record for this book is available from the British Library.

Series on Advances in Mathematics for Applied Sciences — Vol. 95
APPROXIMATION AND REGULARISATION METHODS FOR
OPERATOR-FUNCTIONAL EQUATIONS

Copyright © 2025 by World Scientific Publishing Co. Pte. Ltd.

ISBN 9789819801688 (hardcover)
ISBN 9789819801695 (ebook for institutions)
ISBN 9789819801701 (ebook for individuals)

For any available supplementary material, please visit
https://www.worldscientific.com/worldscibooks/10.1142/14065#t=suppl

Desk Editors: Selas Hamilton/Rosie Williamson/Shi Ying Koe

Typeset by Stallion Press
Email: enquiries@stallionpress.com

Foreword

This book provides innovative insights into modelling complex singular systems as well as theoretical and numerical tools for analysing dynamic systems. These resources can prove valuable for researchers in applied mathematics and engineering across a wide range of disciplines. The primary focus of this book lies in the examination of singular integral equations within Banach function spaces, encompassing both linear and nonlinear operators. Chapter 3 delves into regularisation methods, which are thoroughly explored and involve the transformation of ill-posed problems into well-posed and stable ones through a variety of techniques. This chapter not only outlines traditional approaches for attaining a smooth and stable solution by balancing data fidelity with solution stability but also presents recent research on Lavrentiev regularisation for first-kind equations, non-local continuation of solutions through homotopy analysis, stochastic arithmetic, and the use of block operators. Furthermore, a thorough discussion on bifurcation theory concerning Cauchy problems and integral equations is provided to enhance our comprehension of various intricate phenomena, including the emergence of multiple solutions, branching solutions, periodic orbits, oscillations, and chaos. It also provides insights into critical transitions and potential instabilities within the realm of inverse problems related to the design and control of systems in mechanical and chemical engineering. Chapter 5 explores various application research areas within power systems and engineering, including the magnetic regime in power electronics, biomass gasifier systems, infrared flame tomography, energy storage systems, and battery modelling for new energy. The book presents nonlinear dynamics analysis, its associated numerical methods, and deep reinforcement learning techniques through real-world examples. Consequently, it offers a valuable educational tool for guiding senior undergraduate and graduate students in a range of research projects within modern mathematics, physics, engineering, and related fields, and it also serves as a beneficial reference for researchers in a multitude of application and engineering domains.

Professor Ting Gao
Director of Steklov–Wuhan Institute for Mathematical Exploration,
Huazhong University of Science and Technology, China

About the Authors

Denis Sidorov (DSc, PhD) was born in Irkutsk, Russia, in 1974. He is currently a chair professor at the Harbin Institute of Technology, PRC, and a principal researcher with both the Melentiev Energy Systems Institute of the Siberian Branch of Russian Academy of Sciences and Irkutsk National Research Technical University in Russia. He served as a distinguished guest professor at Hunan University, PRC, and Queen's University Belfast, UK, between 2016 and 2020. Professor Sidorov was elected Chapter Chair of the IEEE Power & Energy Society Russia (Siberia) Chapter between 2018 and 2022. He has authored more than 140 scientific papers and four monographs. His research interests include integral and differential equations, machine learning, wind energy, and inverse problems.

Edixon Rojas (PhD) was born in Venezuela in 1978. He is currently an associate professor with the Department of Mathematics at the Universidad Nacional de Colombia in Bogotá, Colombia. Prior to that, he was an assistant professor at the same department between 2014 and 2018. He received his PhD from the Universidade de Aveiro, Portugal, in 2010. He has authored 50 research papers and delivered more than 20 talks at international conferences. His scientific interests include operator theory, singular and nonlinear integral equations, ordinary differential equations, nonlinear functional analysis, and function spaces.

Alexander Sinitsyn (DSc, PhD) was born in Irkutsk, Russia, in 1961. He is currently a professor at the Universidad Nacional de Colombia in Bogotá, Colombia. He has been a visiting professor at a number of leading research institutions, including Paul Sabatier University (1995, 1996, and 2000), the Department of Mathematics at the Ludwig Maximilian University of Munich (2000), the Erwin Schrödinger International Institute for Mathematics and Physics (2000), the Institute of Mathematics at the Chinese Academy of Sciences (2009 and 2018), and the Technion — Israel Institute of Technology (2009 and 2018). In 2000, Professor Sinitsyn, along with Professor P. Degond and Professor P. Markowich, won the European INTAS grant. His research interests include partial and ordinary differential equations, asymptotics, stability, qualitative properties of Vlasov–Maxwell systems, boundary value problems for semilinear nonlocal elliptic equations, steady-state and nonstationary solutions of Vlasov–Maxwell–Fokker–Planck systems, and applications to earthquake source modelling.

Nikolai Sidorov (DSc, PhD) was born in Irkutsk, Russia, in 1940. He serves as a chair emeritus professor of mathematical analysis and differential equations at Irkutsk State University, Russia. He has been invited to talk at ICIAM congresses and has served as a guest professor at Warwick University, Cambridge University, Chalmers University, Fudan University, Edinburgh University, the Engineering University of Hamburg, and the Banach Mathematical Centre in Warsaw. Professor Sidorov has also been awarded the title of Honoured Scientist of Russia. His scientific interests include branching of solutions of nonlinear equations, the bifurcation theory, singular boundary value problems, regularisation and approximate methods, differential-operator equations, and kinetic systems.

Acknowledgements

We are grateful to the following Professors for their continuous support and motivating discussions: Luis P. Castro, Aoife Foley, Alexander Leonov, Yong Li, Marco Marletta, Robert Plato, Maksim Shishlenin, Valery Sizikov, Maksim Staritsyn, Aleksander Tynda, Anatoly Yagola, Tong Yang and Corresponding Members of the Russian Academy of Sciences Vladislav Pukhnachev and Sergey Kabanikhin.

This work was supported by the Ministry of Science and Higher Education of the Russian Federation, grant number [FZZS-2024-0003] and in part by the National Foreign Experts Program of China, grant number [DL2023161002L] and Beijing Science and Technology Planning Project, grant number [MH20210194].

Much of this book was written at the tables in the library of the Harbin Institute of Technology. We would like to thank Professors Liguo Wang and Fang Liu for their hospitality and Dr. Aliona Dreglea for her invaluable help.

The results presented in this monograph were discussed and prepared during the authors' research visits conducted with support of the following institutions: Universidad Nacional de Colombia, Central South University, City University of Hong Kong, Hong Kong Polytechnic University, Huazhong University of Science and Technology, Hohai University, China University of Geosciences, Queen's University Belfast, and the Isaac Newton Institute for Mathematical Sciences.

Contents

List of Figures

Chapter 1

Introduction

This monograph consists of five parts: this introductory section followed by the four content-focussed chapters, which are unified by a number of key themes: approximate methods, operator theory, regularisation theory, elements of nonlinear functional and numerical analysis, and their wide range of applications in modern science and technology.

Nonlinear dynamical models with parameters can accurately represent complex and chaotic systems, which linear models cannot describe as precisely. Nonlinear models allow us to gain a better understanding of the behaviour and interactions among various variables in a nonlinear system, thereby providing insights that can inform decision-making, predictions, and optimisations. Such dynamical models play a crucial role in the natural sciences, serving as essential tools for analysing and solving complex problems in a range of engineering fields. Parameter-dependent models are commonly used in mathematical modelling across different research areas to tackle both direct and inverse ill-posed problems. The cutting-edge theory of nonlinear differential-operator equations is utilised to model critical processes in fluid dynamics, thermodynamics, and space plasma physics. Models such as the Vlasov–Maxwell systems and Navier–Stokes equations hold particular significance in modern mathematical physics, mechanics, and their applications. The topics discussed in this monograph are not only mathematically intriguing but also increasingly important in electric and thermal engineering, plasma physics, computer tomography, and mechanics, highlighting the need for researchers to acquaint themselves with these contemporary research areas. Examples of the industrial applications of novel dynamical models involving the Volterra equations, the Abel equations, and boundary value problems are included in the concluding part of this monograph.

The fundamentals of bifurcation theory for functional equations stem from the methods developed by A. M. Lyapunov and E. Schmidt in the early 20th century, which were later expanded upon by several renowned mathematicians. For in-depth insights into this topic, interested readers can explore seminal works by A. I. Nekrasov, L. A. Lusternik, M. M. Vainberg, V. A. Trenogin, M. A. Krasnoselsky, B. Buoni, E. N. Dancer, J. Toland, S.-N. Chow, and J. K. Hale, among others. These foundational studies laid the groundwork for modern nonlinear analysis, with significant applications in fluid dynamics and mathematical modelling.

Exploring the branching (or bifurcating) solutions of functional equations is a vital element of analysing such models. The origin of bifurcation theory in relation to functional equations can be traced back to Lyapunov's pioneering work on the equilibrium shapes of rotating liquids. The Lyapunov–Schmidt method, derived from Lyapunov's approach, has since been further developed. Notably, J. H. Poincaré's ideas played a significant role in laying the foundation for bifurcation theory. To establish theorems regarding the presence of bifurcation points, multidimensional extensions of topological and variational methods were introduced by Krasnoselsky, Vainberg, and others, as documented in this book. Renowned mathematicians such as Nekrasov, T. Levi-Chivit, and N. Kochin have conducted in-depth studies on the practical applications of bifurcation theory in mechanics, including convection, wave theory, oscillations, aero-hydro-elasticity, and structural mechanics. The influence of the Lyapunov–Schmidt method and bifurcation theory has extended across a wide range of disciplines, reaching beyond science and engineering into fields such as economics, and thereby broadening their scope.

The Lyapunov–Schmidt method and bifurcation theory also enable us to analyse complex nonlinear systems and understand the behaviour of these systems near critical points, making them essential in power and thermal engineering applications. In these applications, systems often exhibit nonlinear behaviour, leading to dynamic phenomena that elude straightforward understanding through linear analysis. By using the Lyapunov–Schmidt method, engineers can not only identify critical points and stability boundaries but also determine the existence of multiple solutions in nonlinear systems. This is crucial in power and thermal engineering, where system stability, efficiency, and reliability are of utmost importance. Moreover, bifurcation theory enables engineers to predict and control system behaviour near critical points, which is valuable in optimising the performance of power and thermal systems while preventing undesirable outcomes such as oscillations, instabilities, or system failures.

Insights gained by applying these methods help engineers design more robust and efficient power and thermal systems, ensuring their safe and reliable operation.

The groundbreaking contributions of Lyapunov, Poincaré, and Schmidt set the stage for further studies into nonlinear parameter-dependent equations, leading to significant advancements in the natural sciences and engineering over the past century. V. I. Yudovich pioneered the use of symmetry methods in this field, while subsequent works have applied the Lyapunov–Schmidt method, the Conley index theory, and central manifold methods within the framework of group symmetry. These methods have been instrumental in modelling critical processes in plasma physics, fluid dynamics, and thermodynamics through the analysis of nonlinear differential-operator parameter-dependent equations, as well as in industrial applications of materials science, such as fibre melt spinning.

The blow-up theory examines the behaviour of differential-operator equations near special points, particularly bifurcation points, where solutions develop singularities. Analytically, constructing asymptotic solutions in the form of Laurent series and logarithmic-step polynomials has proven to be an effective strategy for understanding the phenomenon of solution blow-up. The irregularity of the problem, particularly when solving systems with irreversible operators as dominant components, has been further explored in works by G. A. Sviridyuk and M. O. Korpusov. Methods for studying solution blow-up in nonlinear partial differential equations have been advanced by researchers such as Korpusov, A. G. Sveshnikov, V. A. Galaktionov, and S. I. Pohožaev. The evolving field of blow-up theory remains a vibrant area of research, with its relevance extending beyond pure mathematics into various physical domains, such as catastrophe theory, thermal and plasma physics, heat and power engineering, fluid mechanics, and numerous other practical applications. In this monograph, the theory of bifurcation analysis is extended to a new class of integral equations with loads, as discussed in Chapter 4.

The theory of bifurcation is closely related to regularisation theory. The term 'ill-posed problem' was first introduced by J. Hadamard (1865–1963). Specifically, according to Hadamard, a problem is considered well posed if a solution exists in some class of functions, the solution is unique in that class, and the solution continuously depends on the input data of the problem (i.e. the initial and boundary conditions, the coefficients of the equation representing the problem, etc.).

Conventional regularisation methods for linear operator equations include Tikhonov regularisation, which introduces a regularisation parameter to balance fidelity with respect to the data and the stability of the solution; truncated singular value decomposition (TSVD), which truncates the singular value decomposition of the operator to stabilise the solution; and total variation regularisation, which penalises the total variation of the solution to achieve a smooth and stable solution. All these methods aim to regularise the linear operator equations by balancing fidelity with respect to the measured data and the stability of the method. For a comprehensive introduction to the theory and numerical methods for the solution of ill-posed problems, readers may refer to the seminal books by Tikhonov *et al.* (1995), Tikhonov *et al.* (1997), Kabanikhin (2012), and Lavrentiev and Savelyev (1995) and the handbook by Sizikov and Verlan (1986). To understand the connections between the classical theories of ill-posed problems and the origins of contemporary machine learning, readers may refer to the book by Vapnik (2006). It is worth noting that, for practical applications, regularisation is not always possible, even when dealing with linear operators, and requires the involvement of (additional) *a priori* knowledge about the nature of the dynamical system being studied.

This monograph builds upon the authors' prior work presented in their earlier monographs (Sidorov *et al.*, 2002; Sidorov, 2024b; Sidorov *et al.*, 2020c) focusing on the foundational theory of dynamical systems governed by operator-functional equations with parameters, including boundary value problems and associated singular integral equations. In addition to exploring the qualitative aspects of dynamical models involving parameters, it delves into practical applications in fields such as power electronics, anaerobic digestion of biomass, and hereditarial dynamics of energy storage.

This monograph is devoted to studying the contemporary aspects of the theory of operator–functional equations, including Volterra operators. In this regard, it goes into more detail about the results presented in the authors' previous works, including Sidorov *et al.* (2021), Sidorov and Dreglea Sidorov (2022, 2023), Sidorov and Sidorov (2021, 2022), Dreglea Sidorov *et al.* (2023), Noeiaghdam *et al.* (2020a, 2020b, 2021), Rafeiro and Rojas (2014), Rojas (2015), Sidorov *et al.* (2019, 2020a), Sizikov and Sidorov (2016), and Muftahov *et al.* (2017). Since this topic is quite extensive, the monograph divides the material into the following four chapters for the convenience of the readers.

Chapter 2: Solvability of Singular Integral Equations on Banach Function Spaces

The first formulation of linear boundary value problems (also called Riemann problems or two-term boundary value problems) for analytic functions was proposed by Riemann (see Riemann, 1867). The theory of singular integral equations, in which the integral is defined in terms of its principal value, emerged almost immediately after the development of the classical theory of integral equations by E. Fredholm in 1903. Singular integral equations were investigated by Hilbert (1912) and Poincaré (1910) while studying two different problems: certain boundary value problems of analytic functions by the former and the theory of tides by the latter. Plemelj (1908) applied further the Cauchy singular integral as a mathematical device for solving boundary value problems.

The complete solution of the Riemann problem was first given by Gakhov (1937, 1990) and Muskhelishvili (1941, 1968). Subsequently, several authors have extensively studied boundary value problems and singular integral equations in classical spaces of integrable functions. Major contributions to the fundamental study of these equations were made by Clancey and Gohberg (1981), Gohberg and Krupnik (1992a, 1992b), Khuskivadze *et al.* (1998), Litvinchuk (2000), Litvinchuk and Spitkovsky (1987), Mikhlin and Prössdorf (1986), Tricomi (1985), Vekua *et al.* (1967), and the references therein.

In the past three decades, several investigations have focused on the study of singular integral equations and boundary value problems in the general setting of variable Lebesgue spaces, $L^{p(\cdot)}$, which is one of the prototypical examples of nonstandard Banach function spaces. The basic properties of boundedness, invertibility, and the Fredholm property of singular integral operators over diverse domains and curves in $L^{p(\cdot)}$ (including weighted versions) as well as the solvability theory of singular integral equations were established by various authors; we refer to Karlovich (2009), Karlovich and Lerner (2005), Karlovich and Spitkovsky (2014), Kokilashvili *et al.* (2007), Kokilashvili and Samko (2003a, 2003b), and their references.

Riemann boundary value problems for analytic functions within the framework of $L^{p(\cdot)}$ were first explored by Kokilashvili *et al.* (2005), while Haseman's problem, the Riemann–Hilbert problem, and the Dirichlet problem, among other kinds of boundary problems, were considered by

Kokilashvili and Paatashvili (2007, 2008, 2009, 2011) and Paatashvili (2010). Their monograph, *Boundary Value Problems for Analytic and Harmonic Functions on Nonstandard Banach Function Spaces* (2012) synthesises much of the developments in this direction.

In the more abstract scheme of Banach functions spaces, singular integral operators and their corresponding equations, with coefficients belonging to different classes of functions, have been studied by Karlovich (1998) (in the case of reflexive rearrangement-invariant spaces), Karlovich and Lerner (2005), and Kokilashvili and Samko (2003a, 2003b), among others. However, despite these developments, the solvability theory of singular integral equations remains far from complete within this general framework.

The aim of Chapter 2 of this monograph is to examine the solvability and Fredholmness of two-term boundary value problems involving analytic functions represented by Cauchy-type integrals considering densities on Banach function spaces over Lyapunov curves, assuming certain conditions. The coefficients for those problems are considered continuous, piecewise continuous, or essentially bounded functions. In each of these cases, since singular integral equations are associated with boundary value problems, the representation of their solutions is used to describe the explicit form of the solutions of the associated equations. For the case of essentially bounded functions, the notion of Wiener–Hopf factorisation is introduced, and then the Simonenko–Fredholm criterion for singular integral operators with factorable functions is established in this context. Moreover, the solvability and explicit representation of the solutions of a class of singular integral equations with shifts are also considered.

The chapter begins with an introduction to Banach function spaces, including key definitions and facts. Here, we introduce the notion of factorisation in $\mathscr{X}(\Gamma)$. Section 2.1 is devoted to the study of the solvability and representation of the solutions for Riemann problems with continuous, piecewise continuous, or essentially bounded factorable coefficients.

In Section 2.2, the Simonenko–Fredholm criterion for singular integral equations with essentially bounded factorable coefficients is proved. As a consequence of this result, we will be able to establish an effective solution for the corresponding singular integral equations using the lateral inverses of the operator.

In Section 2.3, by using the Fredholm criteria for singular integral equations with continuous and piecewise continuous functions proved by Kokilashvili and Samko (2003a), the Simonenko–Fredholm criterion proved in Section 2.2, and the representations of the solutions of the boundary value problems considered in Section 2.1, we aim to show the Fredholm property for the above-mentioned problems and then describe the form of the solutions of the corresponding equation for each class of essentially bounded functions under study.

Section 2.4 deals with a class of singular integral equations with Carleman shifts. For the shift function, we assume one of two behaviours: preserving or reverting the orientation of a curve, Γ. The existence and uniqueness of the eventual solutions of the equation are ensured by the use of projection methods, enabling us to transform the initial equation into a system of equations which can be solved by means of a Riemann boundary value problem technique. Thus, using the results of Section 2.1 and the Sokhotski–Plemelj formulas, we derive the explicit form of the solutions. Furthermore, using the Fredholm criteria mentioned above and the projection methods, which in this case take the form of a non-explicit equivalence relation between operators, the Fredholm property for the associated singular integral operator with a shift is obtained.

In Section 2.4.4, we show that all the assumptions imposed on the abstract Banach function space $\mathscr{X}(\Gamma)$ are, in fact, well-known results in variable Lebesgue spaces and that our results are therefore valid in these spaces.

Chapter 3: Approximation Methods for Linear Operator Equations and Nonlinear Integral Equations

The objective of this chapter is to provide an overview of the authors' recent findings on Lavrentiev regularisation equations of the first kind, the method of non-local continuation of solutions with respect to a parameter (homotopy analysis), and regularisation using stochastic arithmetic or block operators.

Chapter 3 deals with methods for solving series of linear and non-linear operator equations, including Volterra equations of the first kind, applications of the homotopy analysis method, and Adomian decomposition. The chapter is organised as follows. Section 3.1 introduces the perturbation method for the first-kind operator equations. Section 3.2 considers the application of Lavrentiev's α-regulation to solve equations involving Volterra operators with special kernels. The method of non-local continuation of solutions with respect to a parameter (homotopy analysis method) is considered in Section 3.3. Section 3.4 deals with the application of the Adomian decomposition method to Volterra integral equations of the second kind. The novel theory of block operators on normed spaces is introduced in Section 3.5 using the Frobenius formula.

Chapter 4: Loaded Equations and Bifurcation Analysis

The objective of this chapter is to contribute to the bifurcation theory for integral equations and Cauchy problems with loads. This theory is essential for applications due to the following reasons:

(1) *Complex behaviour of dynamic systems*: Bifurcation theory helps in understanding the complex behaviour of dynamic systems, particularly those described by ordinary differential equations and integral equations. By analysing how solutions of Cauchy problems change as parameters vary, bifurcation theory provides insights into the emergence of multiple solutions, branching solutions, periodic orbits, and chaotic behaviour.

(2) *Engineering and physics*: Bifurcation analysis plays a pivotal role in addressing both direct and inverse problems in engineering and physics, enabling engineers and scientists to gain deeper understanding of the behaviour of systems near critical points and predict potential instabilities. This is crucial for designing and controlling systems such as electrical circuits, mechanical systems, and chemical reactors.

(3) *Biological systems*: Bifurcation theory is used in the study of biological systems to understand physiological processes, population dynamics, and ecological interactions. It helps in elucidating how transitions between different states, such as stable and oscillatory behaviours, arise in biological systems.

(4) *Climate science and meteorology*: Bifurcation theory is utilised in the study of stability and transitions within climate and weather patterns. It aids in predicting and understanding the occurrence of phenomena such as abrupt climate change, El Niño events, and atmospheric oscillations, which have direct impacts on distributed generation and forecasting of renewable energy.

(5) *Control and optimisation*: Bifurcation analysis is essential for the control and optimisation of power systems, as it provides crucial information about their stability and performance.

Chapter 4 focuses on the theory and bifurcation analysis of loaded equations. Section 4.1 deals with nonlinear Volterra integral equations with local and Stieltjes-type integral loads. Cauchy problems for nonlinear loaded differential equations with bifurcation parameters are studied in Section 4.2. The theory of linear Fredholm integral equations is introduced in Section 4.3.

Chapter 5: Applications in Electrical Engineering and Automation

This chapter explores the practical applications of approximate and regularised analytical and numerical methods for solving operator-functional equations with parameters in the domains of electric and thermal engineering, including the design of high-energy devices, biogas production, flame tomography, and modelling energy storage systems.

This chapter is organised as follows. Section 5.1 studies the stationary boundary value problem derived from the magnetic (non-)insulated regime in a planar diode. Section 5.2 models the anaerobic digestion process for biogas production using Abel differential equations. Numerical methods and applications of infrared flame tomography of the singular integral equations of Abel type are discussed in Section 5.3. A mathematical model of the nonlinear dynamics of energy storage systems is introduced and validated using a real dataset in Section 5.4.

Chapter 2

Solvability of Singular Integral Equations on Banach Function Spaces

This chapter is devoted to studying the solvability of the equation

$$u(t)\varphi(t) + v(t)(S\varphi)(t) = f(t)$$

posed on a Banach function space (BFS). Here, u and v are certain essentially measurable bounded functions, and $(Sf)(t)$ is the Cauchy singular integral operator along the curve Γ with a finite length ℓ, defined by the formula

$$(Sf)(t) := \frac{1}{\pi i} \text{p.v.} \int_\Gamma \frac{f(\tau)}{\tau - t} d\tau, \quad t = t(s), \quad 0 \leq s \leq \ell,$$

where the integral is understood in the sense of its principal value. Since singular integral equations (SIEs) are related to boundary value problems (BVPs), they are addressed by studying the solvability and Fredholmness of certain two-term BVPs involving analytic functions. These functions are represented by Cauchy-type integrals considering densities on BFSs over Lyapunov curves, assuming certain conditions (see (2.2)–(2.6) in the following).

We start by letting $\Gamma = \{t \in \mathbb{C} : t = t(s),\ 0 \leq s \leq \ell\}$ be an oriented, rectifiable closed simple Lyapunov curve in the complex plane $\mathbb{C}$ with an arc length of s. We denote by D^+ and D^- the bounded and unbounded components of $\mathbb{C} \setminus \Gamma$, respectively. We assume that $z = 0 \in D^+$, and as usual, we consider that Γ is oriented with the natural orientation in the counterclockwise sense. A simple oriented curve Γ in the complex plane is called a Lyapunov curve if the tangent to Γ at each point t exists and forms an angle with the real axis $\theta(t)$, which satisfies the Hölder condition:

$$|\theta(t_1) - \theta(t_2)| \le A|t_1 - t_2|^{\mu}, \quad A > 0, \quad 0 < \mu < 1.$$

By $\mathcal{R}(\Gamma)$, we denote the Banach algebra of the rational functions without poles on Γ, which, as is well known, can be decomposed as $\mathcal{R}(\Gamma) = \mathcal{R}_+(\Gamma) + \mathcal{R}_-(\Gamma)$, where $\mathcal{R}_{\pm}(\Gamma)$ denotes the set of all functions on $\mathcal{R}(\Gamma)$ with poles outside of $D^{\pm}$. The continuous functions, smooth functions, and essentially measurable bounded functions on Γ, endowed with the essential supremum norm $\| \cdot \|_{\infty}$, are denoted by $C(\Gamma)$, $C^{\infty}(\Gamma)$, and $L^{\infty}(\Gamma)$, respectively. By $L^p(\Gamma)$, for $p \in [1, \infty)$, we denote the usual Banach space of all Lebesgue-measurable complex-valued functions on Γ with an absolutely integrable pth power.

Let (Ω, μ) be a non-atomic σ-finite measure space, i.e. a measure space with a non-atomic σ-finite measure μ given on a σ-algebra of the subsets of Ω. The set of all Lebesgue-measurable complex-valued functions on Ω is denoted by $\mathcal{M}$. Let $\mathcal{M}^+$ be the subset of functions in $\mathcal{M}$ whose values lie in $[0, \infty]$. The characteristic function of a measurable set $E \subset \Omega$ is denoted by χ_E, and the Lebesgue measure of E is denoted by $|E|$.

Definition 2.0.1 (Bennett and Sharpley, 1988, Chapter 1, Definition 1.1). A mapping $\rho : \mathcal{M}^+ \to [0, \infty]$ is called a *Banach function norm* if, for all functions f, g, f_n $(n \in \mathbb{N})$ in $\mathcal{M}^+$, for all constants $a \ge 0$, and for all measurable subsets E of Ω, the following properties hold:

(A1) $\rho(f) = 0 \Leftrightarrow f = 0$ a.e., $\rho(af) = a\rho(f)$, $\rho(f + g) \le \rho(f) + \rho(g)$;

(A2) $0 \le g \le f$ a.e. $\Rightarrow \rho(g) \le \rho(f)$ (the lattice property);

(A3) $0 \le f_n \uparrow f$ a.e. $\Rightarrow \rho(f_n) \uparrow \rho(f)$ (the Fatou property);

(A4) $|E| < \infty \Rightarrow \rho(\chi_E) < \infty$;

(A5) $|E| < \infty \Rightarrow \int_E f(x) \, dx \le C_E \rho(f)$,

with $C_E \in (0, \infty)$ a constant which may depend on E and ρ but is independent of f.

Here, functions differing only on a set of measure zero are identified. The set $\mathscr{X}(\Omega)$ of all functions $f \in \mathcal{M}$ for which $\rho(|f|) < \infty$ is called a BFS. For each $f \in \mathscr{X}(\Omega)$, the norm of f is defined by

$$\|f\|_{\mathscr{X}(\Omega)} := \rho(|f|).$$

The set $\mathscr{X}(\Omega)$ under natural linear space operations and under this norm becomes a Banach space (see Bennett and Sharpley, 1988, Chapter 1, Theorems 1.4 and 1.6).

If ρ is a Banach function norm, its associate norm ρ' is defined on $\mathcal{M}^+$ by

$$\rho'(g) := \sup\left\{ \int_\Omega f(x)g(x)\,dx \;:\; f \in \mathcal{M}^+, \; \rho(f) \leq 1 \right\}, \quad g \in \mathcal{M}^+,$$

which is a Banach function norm itself (Bennett and Sharpley, 1988, Chapter 1, Theorem 2.2). The BFS $\mathscr{X}'(\Omega)$ determined by the Banach function norm ρ' is called the associate space (or Köthe dual) of $\mathscr{X}(\Omega)$.

Lemma 2.0.1 (Hölder's inequality, see Bennett and Sharpley, 1988, Chapter 1, Theorem 2.4). *Let $\mathscr{X}(\Omega)$ be a BFS with an associate space $\mathscr{X}'(\Omega)$. If $f \in \mathscr{X}(\Omega)$ and $g \in \mathscr{X}'(\Omega)$, then fg is summable and*

$$\int_\Omega |fg|d\mu \leq \|f\|_{\mathscr{X}(\Omega)}\|g\|_{\mathscr{X}'(\Omega)}. \tag{2.1}$$

Let Σ denote the collection of all subsets of Ω of finite measure, where any two such subsets which differ by a set of μ-measure zero are identified. With the distance

$$d(E,F) := \int_\Omega |\chi_E - \chi_F|d\mu, \quad E,F \in \Sigma,$$

(Σ, d) is a complete metric space. A measure μ is said to be separable if the corresponding metric space (Σ, d) is separable.

Lemma 2.0.2 (Bennett and Sharpley, 1988, Chapter 1, Corollaries 4.3–4.5). *Let μ be a separable measure.*

(a) *A BFS $\mathscr{X}(\Omega)$ is separable if and only if its associate space $\mathscr{X}'(\Omega)$ is canonically isometrically isomorphic to the dual space $\mathscr{X}^*(\Omega)$ of $\mathscr{X}(\Omega)$.*
(b) *A BFS $\mathscr{X}(\Omega)$ is reflexive if and only if both $\mathscr{X}(\Omega)$ and its associate space $\mathscr{X}'(\Omega)$ are separable.*

We consider $\mathscr{X}(\Gamma)$ to be a BFS over a closed simple Lyapunov curve Γ satisfying the following conditions:

$$C(\Gamma) \subset \mathscr{X}(\Gamma) \subset L^1(\Gamma). \tag{2.2}$$

$$\|af\|_{\mathscr{X}(\Gamma)} \leq \sup_{t\in\Gamma} |a(t)| \cdot \|f\|_{\mathscr{X}(\Gamma)}, \quad \text{for } a \in L^\infty(\Gamma). \tag{2.3}$$

$$\text{The operator } S \text{ is bounded in } \mathscr{X}(\Gamma). \tag{2.4}$$

$$\mathscr{X}(\Gamma) \text{ is reflexive.} \tag{2.5}$$

$$C^\infty(\Gamma) \text{ is dense in } \mathscr{X}(\Gamma). \tag{2.6}$$

The boundedness of the adjoint operator S^* in the dual space $\mathscr{X}^*(\Gamma)$ is given in the following result.

Lemma 2.0.3. *Let the operator S be bounded in the space $\mathscr{X}(\Gamma)$. Then, the operator S^* adjoint to the operator S is connected with the operator S in the dual space $\mathscr{X}^*(\Gamma)$ via the equality*

$$S^* = -HSH,$$

where H is the operator defined in the space $\mathscr{X}^*(\Gamma)$ by the formula $(H\varphi)(t) := \overline{h(t)\varphi(t)}$, with $h(t) = \exp(i\Theta(t))$ and $\Theta(t)$ being the angle of inclination of the curve Γ at point t to the positive direction of the real axis.

Proof. Since $\mathscr{X}(\Gamma)$ is reflexive, then from Lemma 2.0.2, $\mathscr{X}(\Gamma)$ and $\mathscr{X}'(\Gamma)$ are separable. Furthermore, the dual space $\mathscr{X}^*(\Gamma)$ can be identified with the associate space $\mathscr{X}'(\Gamma)$ (see also Lemma 1.2 in Karlovich, 1998). That is, the general form of a linear functional on $\mathscr{X}(\Gamma)$ is given by

$$f(u) = (u, v) = \int_\Gamma u(t)\overline{v(t)}|dt|, \quad \text{where } u \in \mathscr{X}(\Gamma),\ v \in \mathscr{X}'(\Gamma).$$

Let $\phi, \psi \in \mathcal{R}(\Gamma)$. Then, from Cauchy's theorem, it follows that

$$\int_\Gamma \psi(t)(S\phi)(t)dt = -\int_\Gamma \phi(t)(S\psi)(t)dt.$$

Hence,

$$(\phi, S^*\psi) = (S\phi, \psi) = \int_\Gamma (S\phi)(t)\overline{\psi(t)}|dt| = \int_\Gamma (S\phi)(t)\overline{\psi(t)}h(t)dt$$

$$= -\int_\Gamma \phi(t)(S\overline{h\psi})(t)dt = -\int_\Gamma \phi(t)\overline{(HSH\psi)(t)}|dt| = -(\phi, HSH\psi).$$

$$(2.7)$$

Since $\mathscr{X}(\Gamma)$ is reflexive, $\mathscr{X}(\Gamma)$ and $\mathscr{X}'(\Gamma) = \mathscr{X}^*(\Gamma)$ are separable, and by (2.6), $\mathcal{R}(\Gamma)$ is dense in $\mathscr{X}(\Gamma)$ and $\mathscr{X}'(\Gamma)$. Then, from (2.4) and (2.7), we conclude that $S^* = -HSH$. $\square$

On the other hand, from Mikhlin and Prössdorf (1986, Chapter I, Corollary 1.2), we have that $(S^2r)(t) = r(t)$, for all $r \in \mathcal{R}(\Gamma)$. Since $\mathcal{R}(\Gamma)$ is dense in $C^\infty(\Gamma)$ and by assumption (2.6) $C^\infty(\Gamma)$ is dense in $\mathscr{X}(\Gamma)$, we conclude that $S^2 = I$ in $\mathscr{X}(\Gamma)$. Hence, from (2.4) and (2.6), the operators

$$P_\pm := \frac{1}{2}(I \pm S)$$

define bounded complementary projections in the space $\mathscr{X}(\Gamma)$. Thus, we define the subspaces

$$\mathscr{X}_+(\Gamma) := P_+\mathscr{X}(\Gamma), \quad \overset{\circ}{\mathscr{X}}_-(\Gamma) := P_-\mathscr{X}(\Gamma),$$

$$\mathscr{X}_-(\Gamma) := \overset{\circ}{\mathscr{X}}_-(\Gamma) + \mathbb{C}.$$

Putting

$$L^1_+(\Gamma) := \left\{ f \in L^1(\Gamma) \; : \; \int_\Gamma f(\tau)\tau^n d\tau = 0, \text{ for } n \ge 0 \right\},$$

$$\overset{\circ}{L^1}_-(\Gamma) := \left\{ f \in L^1(\Gamma) \; : \; \int_\Gamma f(\tau)\tau^{-n} d\tau = 0, \text{ for } n \ge 1 \right\},$$

$$L^1_-(\Gamma) := \overset{\circ}{L^1}_-(\Gamma) + \mathbb{C},$$

the following lemma holds.

Lemma 2.0.4. *Let Γ be a Lyapunov curve and $\mathscr{X}(\Gamma)$ be a BFS satisfying (2.2)–(2.6). Then:*

(a) *$\mathscr{X}_+(\Gamma) = L^1_+(\Gamma) \cap \mathscr{X}(\Gamma)$, $\overset{\circ}{\mathscr{X}}_-(\Gamma) = \overset{\circ}{L^1}_-(\Gamma) \cap \mathscr{X}(\Gamma)$, and $\mathscr{X}_-(\Gamma) = L^1_-(\Gamma) \cap \mathscr{X}(\Gamma)$.*
(b) *If $f \in \mathscr{X}_\pm(\Gamma)$, $g \in \mathscr{X}'_\pm(\Gamma)$, then $fg \in L^1_\pm(\Gamma)$. Moreover, if $f \in \mathscr{X}_-(\Gamma)$ and $g \in \overset{\circ}{\mathscr{X}'}_-(\Gamma)$ or $f \in \overset{\circ}{\mathscr{X}}_-(\Gamma)$ and $g \in \mathscr{X}'_-(\Gamma)$, then $fg \in \overset{\circ}{L^1}_-(\Gamma)$.*

Proof. The proof of (a) follows from Mikhlin and Prössdorf (1986, Chapter II, Theorem 1.1) and assumptions (2.2) and (2.4), taking into consideration the decomposition $\mathcal{R}(\Gamma) = \mathcal{R}_+(\Gamma) + \mathcal{R}_-(\Gamma)$. The proof of (b) runs analogous to the proof of Lemma 6.11 in Böttcher and Karlovich (1997), from the denseness of $\mathcal{R}(\Gamma)$ on $C(\Gamma)$, assumption (2.6), and the Hölder inequality. ☐

Now, we can introduce the factorisation of an invertible, essentially measurable bounded function a in Γ, ($a \in \mathcal{G}(L^\infty(\Gamma))$), where $a \in \mathcal{G}(L^\infty(\Gamma))$ if ess $\inf_{t \in \Gamma} |a(t)| > 0$. Let Γ be a Lyapunov curve and $\mathscr{X}(\Gamma)$ be a BFS satisfying (2.2)–(2.6). We say that a function $a \in \mathcal{G}(L^\infty(\Gamma))$ admits a *factorisation in $\mathscr{X}(\Gamma)$* if it can be written in the form

$$a(t) = a_-(t)t^{\aleph}a_+(t), \quad \text{a.e. on } \Gamma, \tag{2.8}$$

where $\aleph \in \mathbb{Z}$.

(i) $a_- \in \mathscr{X}_-(\Gamma)$, $a_-^{-1} \in \mathscr{X}'_-(\Gamma)$, $a_+ \in \mathscr{X}'_+(\Gamma)$, $a_+^{-1} \in \mathscr{X}_+(\Gamma)$.

(ii) The operator $a_+^{-1} S a_+ I$ is bounded in $\mathscr{X}(\Gamma)$.

The integer $\aleph$ is called the index of the function a and is denoted by $\operatorname{ind} a$. It can be proved that the number $\aleph$ is uniquely determined.

2.1   Binomial BVPs

Let Ψ be an analytic function of Cauchy-type integral with a non-tangential limit $\varphi \in \mathscr{X}(\Gamma)$:

$$\Psi(z) = \frac{1}{2\pi i} \int_\Gamma \frac{\varphi(\tau)}{\tau - z} d\tau, \quad z \notin \Gamma,$$

with boundary values $\Psi^+(t)$ ($\Psi^-(t)$), where $z \to t$, $t \in \Gamma$, and $z \in D^+$ ($t \in \Gamma$, $z \in D^-$). A Riemann problem, or a binomial BVP, for analytic functions represented by Cauchy-type integrals with densities on the space $\mathscr{X}(\Gamma)$ is stated in the following way: find the functions $\Psi^+(t)$ and $\Psi^-(t)$ analytic in D^+ and D^-, respectively, vanishing at infinity, and satisfying the condition

$$\Psi^+(t) = G(t)\Psi^-(t) + g(t), \tag{2.9}$$

or, equivalently, the associate problem

$$\Psi^-(t) + G(t)\Psi^+(t) = g(t) \tag{2.10}$$

imposed on their boundary values on the contour Γ. The pair $\{\Psi^-, \Psi^+\}$ is referred to as the solution of problem (2.9) (or (2.10)). The subspace of the solutions of the homogeneous problem is called its kernel, and the subspace of the functions g, for which the inhomogeneous problem is solvable, is said to be its image. The dimension α of the first and the codimension (in $\mathscr{X}(\Gamma)$) β of the closure of the second are called the defect numbers of the BVP. If at least one of the numbers α and β is finite, the difference $\alpha - \beta$ is referred to as its index. The problem (2.9), or (2.10), is said to be normally solvable if its image is closed, whereas it is called Fredholm if it is normally solvable and has a finite index.

On the other hand, according to the Sokhotski–Plemelj formulas, the boundary values of Ψ are expressed by

$$\Psi^+(t) = \frac{1}{2}[(I\varphi)(t) + (S\varphi)(t)], \quad \Psi^-(t) = \frac{1}{2}[(-I\varphi)(t) + (S\varphi)(t)].$$

Therefore, $\Psi^+(t) - \Psi^-(t) = \varphi(t)$ and $\Psi^+(t) + \Psi^-(t) = (S\varphi)(t)$, which allows us to reduce the SIE

$$(a(t)P_+ + b(t)P_-)\varphi(t) = g(t), \quad b \in \mathcal{G}(L^\infty(\Gamma)),$$

or, equivalently,

$$(a(t)b^{-1}(t)P_+ + P_-)\varphi(t) = b^{-1}(t)g(t), \qquad (2.11)$$

to the Riemann BVP

$$\Psi^-(t) + (-a(t)b^{-1}(t))\Psi^+(t) = -b^{-1}(t)g(t). \qquad (2.12)$$

Analogously, the equation

$$(a(t)P_+ + b(t)P_-)\varphi(t) = g(t), \quad a \in \mathcal{G}(L^\infty(\Gamma))$$

$$(P_+ + b(t)a^{-1}(t)P_-)\varphi(t) = a^{-1}(t)g(t) \qquad (2.13)$$

is reduced to the Riemann BVP

$$\Psi^+(t) = (b(t)a^{-1}(t))\Psi^-(t) + a^{-1}(t)g(t) \qquad (2.14)$$

for a piecewise analytic function $\{\Psi^+(z), \Psi^-(z)\}$ vanishing at the point $z = \infty$. That is, equation (2.11) and problem (2.12), as well as equation (2.13) and problem (2.14), with the additional condition $\Psi^-(\infty) = 0$, are equivalent. This means there exists a one-to-one correspondence between the solutions of problem (2.12) (resp., (2.14)) and the solutions of equation (2.11) (resp., (2.13)).

For a simply connected domain D, bounded by a rectifiable curve Γ, we denote by $E^\delta(D)$, $\delta > 0$, the Smirnov class of functions $\Psi(z)$ in D, for which

$$\sup_r \int_{\Gamma_r} |\Psi(z)|^\delta |dz| < \infty,$$

where Γ_r is the image of $\gamma_r = \{z : |z| = r\}$ under the conformal mapping of $U = \{z : |z| < 1\}$ onto D. When D is an infinite domain, then the conformal mapping implies transforming zero into infinity. A function $\Psi \in E^\delta(D)$ possesses almost everywhere angular boundary values on Γ, and the boundary function belongs to $L^\delta(\Gamma)$.

Now, let us introduce the notation

$$\mathscr{E}(D) = \Big\{ \Psi(z) \; : \; \Psi(z) = (K\varphi)(z)$$

$$= \frac{1}{2\pi} \int_\Gamma \frac{\varphi(\tau)}{\tau - z} d\tau, \; z \notin \Gamma \text{ with } \varphi \in \mathscr{X}(\Gamma) \Big\}.$$

As is known, $E^1(D)$ coincides with the class of analytic functions represented by Cauchy integrals, for the function $\Psi(z)$, which is analytic on the plane cutting along the closed curve Γ and belongs to $E^1(D^{\pm})$. Then,

$$\Psi(z) = K(\Psi^+ - \Psi^-)(z),$$

which, with the inclusion of $\mathscr{X}(\Gamma) \subset L^1(\Gamma)$, given by assumption (2.2), allows us to define the following subsets of $\mathscr{E}(D)$:

$$\mathscr{E}^1(D^{\pm}) = \{\varphi \in E^1(D^{\pm}) \; : \; \varphi \text{ has definite limiting values on } \mathscr{X}(\Gamma)\},$$

where $\mathscr{E}^1_+(D^+)$ denotes the set of the analytic functions on D^+ with definite limiting on $\mathscr{X}_+(\Gamma)$. $\mathscr{E}^1_-(D^-)$ is referred to as the set of analytic functions on D^- vanishing at infinity with boundary value on $\overset{\circ}{\mathscr{X}}_-(\Gamma)$. Finally, we set $\mathscr{L}^1(\Gamma) := L^1_+(\Gamma) \dotplus \overset{\circ}{L}^1_-(\Gamma)$.

Since the BVP is posed in $\mathscr{X}(\Gamma)$, we seek solutions $\{\varphi^+, \varphi^-\}$ represented by integrals of Cauchy type, i.e.

$$\varphi^{\pm}(z) = \frac{1}{\pi i} \int_{\Gamma} \frac{\varphi^{\pm}(\zeta)}{\zeta - z} d\zeta, \quad \text{hold in } D^{\pm},$$

with non-tangential limits a.e. on $\mathscr{X}(\Gamma)$. Then, the solutions are such that $\varphi^+ \in \mathscr{E}^1(D^+)$ and $\varphi^- \in \mathscr{E}^1(D^-)$ for the BVP with continuous coefficients G. In the case of essentially bounded coefficients G admitting a factorisation (2.8), the solutions $\varphi^{\pm}$ represented by the integrals of Cauchy type have non-tangential limits a.e. on $\mathscr{X}_+(\Gamma)$ and $\overset{\circ}{\mathscr{X}}_-(\Gamma)$, respectively. In this case, the solutions are such that $\varphi^+ \in \mathscr{E}^1_+(D^+)$ and $\varphi^- \in \overset{\circ}{\mathscr{E}}^1_-(D^-)$.

2.1.1   *Continuous coefficients*

We now aim to study the solvability of problem (2.9) for G, a non-vanishing continuous function on Γ with index $\aleph = \frac{1}{2\pi}[\arg G(t)]_{\Gamma}$, and the function $g \in \mathscr{X}(\Gamma)$. To attain this objective, we first establish the following auxiliary result.

Proposition 2.1.1. *Let Γ be a closed curve and $\mathscr{X}(\Gamma)$ be a BFS satisfying (2.2)–(2.4) and (2.6). Assume that $z_0 \in D^+$. Then, there exists an integer $k \geq 0$ such that*

$$\exp\{(K\varphi)(z)\} =: X(z) \in \mathscr{E}^1(D^+) \quad \text{and} \quad \frac{X(z) - 1}{(z - z_0)^k} \in \mathscr{E}^1(D^-).$$

Proof. Let $\delta > 0$ and Γ_r be the image of $\gamma_r = \{z \; : \; |z| = r\}$, $r < 1$, under the conformal mapping of $U = \{z \; : \; |z| < 1\}$ onto D^+. We have

$$\int_{\Gamma_r} |X(z)|^{\delta} |dz| \leq \int_{\Gamma_r} \sum_{n=0}^{\infty} \frac{1}{n!} |\delta \Psi(z)|^n |dz|, \quad \text{where} \quad \Psi(z) = \frac{1}{2\pi} \int_{\Gamma} \frac{\varphi(\tau)}{\tau - z} d\tau.$$
(2.15)

Since $\Psi(z) \in E^1(D^+)$, it is known that

$$\int_{\Gamma_r} |\Psi(z)|^n |dz| \leq \int_{\Gamma} |\Psi^+(t)|^n |dt|.$$

From (2.15), we obtain

$$\int_{\Gamma_r} |X(z)|^{\delta} |dz| \leq \sum_{n=0}^{\infty} \int_{\Gamma} |\delta \Psi^+(t)|^n |dt| \leq \sum_{n=0}^{\infty} \frac{1}{n!} \int_{\Gamma} \left| \frac{\delta \varphi(t)}{2} + \frac{\delta}{2} (S\varphi)(t) \right|^n |dt|$$

$$\leq \sum_{n=0}^{\infty} \frac{1}{n!} \int_{\Gamma} |\delta \varphi(t)|^n |dt| + \sum_{n=0}^{\infty} \frac{1}{n!} \int_{\Gamma} |\delta(S\varphi)(t)|^n |dt|$$

$$\leq \ell e^{\delta M} + \sum_{n=0}^{\infty} \frac{1}{n!} \int_{\Gamma} |\delta(S\varphi)(t)|^n |dt|,$$

where $M = \sup_{t \in \Gamma} |\varphi(t)|$. Note that the series $\sum_{n=0}^{\infty} \frac{1}{n!} \int_{\Gamma} |\delta(S\varphi)(t)|^n |dt|$ converges if $\delta \leq 1$. The case of even n was shown above. This proves that $X(z) \in E^{\delta}(D^+)$ when $\delta \leq 1$.

In the case of D^-, it is necessary to consider two cases: $0 < r < r_0$ and $r_0 < r < 1$, for some fixed r_0. The required inequalities are obtained by means of choosing the number $k > [\frac{1}{\delta}]$ and proceeding as before.

Now, note that

$$\int_{\Gamma} |\Psi^{\pm}(t)| |dt| = \int_{\Gamma} \left| e^{\pm \frac{\delta \varphi(t)}{2}} \right| \left| e^{\frac{\delta}{2}(S\varphi)(t)} \right| |dt|$$

$$\leq e^{\frac{\delta M}{2}} \sum_{n=0}^{\infty} \frac{1}{n!} \int_{\Gamma} \left| \frac{\delta}{2}(S\varphi)(t) \right|^n |dt|.$$

As before, we can prove that this series converges if $\delta \leq 2$.

Finally, we apply the following Smirnov's theorem: *let $\Psi \in E^{\gamma_1}(D)$ and $\Psi^+ \in L^{\gamma_2}(\Gamma)$, where $\gamma_2 > \gamma_1$. Then, $\Psi \in E^{\gamma_2}(D)$.* In our case, $X(z) \in E^{\delta}(D^+)$, $0 < \delta < 1$, and assumption (2.2) gives us that $\Psi^+ \in L^1(\Gamma)$. Then, we have $X(z) \in E^1(D^+)$ and $\frac{X(z)-1}{(z-z_0)^k} \in E^1(D^-)$. Furthermore, when we consider analytic functions with non-tangential limits on $\mathscr{X}(\Gamma)$, or, more precisely, $\Psi^+ \in \mathscr{X}(\Gamma)$, then we conclude that $X(z) \in \mathscr{E}^1(D^+)$ and $\frac{X(z)-1}{(z-z_0)^k} \in \mathscr{E}^1(D^-)$. $\qquad \square$

Theorem 2.1.2. *Let Γ be a Lyapunov curve and $\mathscr{X}(\Gamma)$ be a BFS satisfying (2.2)–(2.4) and (2.6). Assume that $G \in C(\Gamma)$ and $G(t) \neq 0$, $t \in \Gamma$. Then, for problem (2.9), the following statements hold:*

(a) *For $\aleph \geq 0$, problem (2.9) is unconditionally solvable in the class $\mathscr{E}(D)$, and all its solutions are given by*

$$\Psi(z) = \frac{X(z)}{2\pi i} \int_\Gamma \frac{g(\tau)}{X^+(\tau)} \frac{d\tau}{\tau - z} + X(z)\rho(z), \qquad (2.16)$$

with

$$X(z) = \begin{cases} \exp h(z), & z \in D^+, \\ (z - z_0)^\aleph \exp h(z), & z \in D^-, \quad z_0 \in D^+, \end{cases}$$

where

$$h(z) = K\left(\ln G(t)(t - z_0)^\aleph\right)(z)$$

and ρ is an arbitrary polynomial of degree $\aleph - 1$.

(b) *For $\aleph < 0$, problem (2.9) is solvable in this class if and only if*

$$\int_\Gamma \frac{g(\tau)\tau^\kappa}{X^+(\tau)} d\tau = 0, \quad \kappa = 0, 1, \ldots, |\aleph| - 1, \qquad (2.17)$$

and under these conditions, problem (2.9) has a unique solution, given by (2.16) with $\rho \equiv 0$.

Proof. Let us consider first the case of $\aleph = 0$. We choose a rational function $\widetilde{G}(t)$ such that

$$\sup_{t \in \Gamma} \left| \frac{G(t)}{\widetilde{G}(t)} - 1 \right| < \frac{1}{2}\left(1 + \|S\|_{\mathscr{X}(\Gamma)}\right)^{-1}.$$

Clearly, $\operatorname{ind} \widetilde{G} = 0$; therefore, $\widetilde{X}(z) = \exp(K(\ln \widetilde{G}))(z)$ is continuous in the domains $D^\pm$. Since $\Psi(z) = (K(\Psi^+ - \Psi^-))(z)$, we have

$$\left(\frac{\Psi}{\widetilde{X}}\right)^+ + \frac{G}{\widetilde{G}}\left(\frac{\Psi}{\widetilde{X}}\right)^- = \frac{g}{\widetilde{X}^+}. \qquad (2.18)$$

Note that $\frac{\Psi}{\widetilde{X}} \in \mathscr{E}(D)$. In fact, because $\Psi \in E^1(D^\pm)$ and $\frac{1}{\widetilde{X}}$ is bounded, $\frac{\Psi}{\widetilde{X}} \in E^1(D^\pm)$ and therefore

$$\frac{\Psi}{\widetilde{X}} = K\left(\frac{\Psi^+}{\widetilde{X}^+} - \frac{\Psi^-}{\widetilde{X}^-}\right).$$

From the Sokhotski–Plemelj formulas, it follows that $\Psi^+ \in \mathscr{X}(\Gamma)$ and hence $\frac{\Psi}{\widetilde{X}} \in \mathscr{E}(D)$. Let

$$\frac{\Psi(z)}{\widetilde{X}(z)} = (K\psi)(z), \quad \psi \in \mathscr{X}(\Gamma).$$

Then, equality (2.18) yields

$$\psi(t) = \left(\frac{G(t)}{\widetilde{G}(t)} - 1\right)\left(-\frac{1}{2}\psi(t) + \frac{1}{2}(S\psi)(t)\right) + \frac{g(t)}{\widetilde{X}(t)}, \tag{2.19}$$

which means the function ψ is a solution of the equation $\psi = K\psi$ in the space $\mathscr{X}(\Gamma)$, where K is a contractive operator. Therefore, equation (2.19), and consequently problem (2.9), has a unique solution in $\mathscr{E}(D)$. Now, we attempt to construct such a solution.

Let

$$X(z) = \exp\left(K(\ln G)(z)\right).$$

Since $\aleph = 0$,

$$\ln G(t) = \ln|G(t)| + i\arg G(t)$$

is a continuous function, and from Proposition 2.1.1,

$$\frac{1}{X(z)} - 1 \in \mathscr{E}^1(D^\pm).$$

If Ψ is a solution to problem (2.9), then $\Psi \in \mathscr{E}(D)$ and therefore $\Psi \in E^1(D^\pm)$; moreover, $\frac{\Psi}{X} \in \mathscr{E}(D)$. Also,

$$\left(\frac{\Psi}{X}\right)^+ - \left(\frac{\Psi}{X}\right)^- = \frac{g}{X^+}.$$

Since this problem has a unique solution in $\mathscr{E}(D)$, the function

$$\Psi(z) = X(z)K\left(\frac{g}{X^+}\right)(z)$$

is a solution to (2.9) in the class $\mathscr{E}(D)$.

Now let $\aleph > 0$. We choose a point $z_0 \in D^+$ and rewrite problem (2.9) as

$$\Psi^+(t) = G_1(t)(t - z_0)^\aleph \Psi^-(t) + g(t),$$

where $G_1(t) = (t - z_0)^{-\aleph}G(t)$ is a continuous function with index zero. Let

$$F(z) = \begin{cases} \Psi(z), & z \in D^+, \\ (z - z_0)^\aleph \Psi(z), & z \in D^-. \end{cases} \tag{2.20}$$

For $F(z)$, there exists a polynomial $\rho(z)$ of degree $\aleph - 1$ such that

$$\Xi(z) = F(z) - \rho(z) \in E^1(D^-). \tag{2.21}$$

Then, $\Xi(z) = K(\Xi^+ - \Xi^-)(z)$. However,

$$\Xi^+(t) - \Xi^-(t) = F^+(t) - F^-(t) = \Psi(t) - (t - z_0)^\aleph \Psi^-(t) \in \mathscr{X}(\Gamma);$$

thus, $\Xi \in \mathscr{E}(D)$. Besides,

$$\Xi^+(t) = G_1(t)\Xi^-(t) + g_1(t),$$

where $g_1(t) = g(t) - \rho(t) + G_1(t)\rho(t)$. Since $\mathrm{ind}\,G_1 = 0$, from the previous part,

$$\Xi(z) = X_1(z)K\left(\frac{g_1}{X_1^+}\right)(z), \quad X_1(z) = \exp\left(K(\ln G_1)(z)\right).$$

On the other hand,

$$K\left(\frac{g_1}{X_1^+}\right)(z) = K\left(\frac{g}{X_1^+}\right)(z) - \frac{1}{2\pi i}\int_\Gamma \frac{\rho(t)}{X_1^+(t)}\frac{dt}{t-z} + \frac{1}{2\pi i}\int_\Gamma \frac{\rho(t)}{X_1^-(t)}\frac{dt}{t-z}.$$

However,

$$\frac{1}{2\pi i}\int_\Gamma \frac{\rho(t)}{X_1^+(t)}\frac{dt}{t-z} = \begin{cases} \dfrac{\rho(z)}{X_1(z)}, & z \in D^+, \\ 0, & z \in D^-, \end{cases}$$

and

$$\frac{1}{2\pi i}\int_\Gamma \frac{\rho(t)}{X_1^-(t)}\frac{dt}{t-z} = \frac{1}{2\pi i}\int_\Gamma \left[\frac{\rho(t)}{X_1^-(t)} - \rho(t)\right]\frac{dt}{t-z} + \frac{1}{2\pi i}\int_\Gamma \frac{\rho(t)}{t-z}dt$$

$$= \begin{cases} \rho(z), & z \in D^+, \\ -\dfrac{\rho(z)}{X_1(z)} + \rho(z), & z \in D^-. \end{cases}$$

Thus,

$$\Xi(z) = X_1(z)K\left(\frac{g_1}{X_1^+}\right)(z) = X_1(z)K\left(\frac{g}{X_1^+}\right)(z) + X_1(z)\rho(z) - \rho(z).$$

From (2.20) and (2.21), we arrive at formula (2.16).

It can be verified that, in the solution provided for problem (2.9), the arbitrary polynomial ρ does not depend on the choice of the point z_0.

Finally, for the case $\aleph < 0$, the function given by (2.20) belongs to $\mathscr{E}(D)$. Moreover, $F^+ = G_1 F^- + g$. Whence,

$$F(z) = X_1(z)K\left(\frac{g}{X_1^+}\right)(z).$$

Cauchy's formula (Golozin, 1969, Chapter X, Section 4, Theorem 1) states that, for any analytic function ϕ with non-tangential limits a.e. on the contour Γ representable by an integral of Cauchy type in D, it is necessary and sufficient that

$$\int_\Gamma \phi(\zeta)\zeta^n d\zeta = 0, \quad n = 0, 1, \ldots.$$

Therefore, the condition $\Psi(z) = (z - z_0)^{-\aleph}F(z) \in E^1(D^-)$ is fulfilled if and only if condition (2.17) is satisfied. $\square$

2.1.2 Piecewise continuous coefficients

In this section, we study the solvability of the two-term BVP (2.9) having piecewise continuous functions as coefficients. For this, first we introduce two necessary axioms on $\mathscr{X}(\Gamma)$ as well as some auxiliary results, which were proved by Kokilashvili and Samko (2003a).

Axiom 1. For the space $\mathscr{X}(\Gamma)$, there exist two functions α and β, $0 < \alpha(t)$, $\beta(t) < 1$, such that

$$|t - t_0|^{\gamma(t_0)} S|t - t_0|^{-\gamma(t_0)} I, \quad t_0 \in \Gamma$$

is bounded in the space $\mathscr{X}(\Gamma)$, for all $\gamma(t_0)$, and such that

$$-\alpha(t_0) < \gamma(t_0) < 1 - \beta(t_0),$$

and is unbounded in $\mathscr{X}(\Gamma)$ if $\gamma(t_0) \notin (-\alpha(t_0), 1 - \beta(t_0))$.

The functions α and β are called the *index functions* of the space $\mathscr{X}(\Gamma)$.

Axiom 2. For any $\gamma < 1 - \beta(t_0)$, the embedding $\mathscr{X}(\Gamma, |t - t_0|^\gamma) \subset L^1(\Gamma)$ is valid and $C^\infty(\Gamma)$ is dense in $\mathscr{X}(\Gamma, |t - t_0|^\gamma)$, regardless of $t_0 \in \Gamma$.

From Axiom 1, the following result holds.

Lemma 2.1.3. *Let the space $\mathscr{X}(\Gamma)$ satisfy conditions (2.2)–(2.3) and $t_1, \ldots, t_n \in \Gamma$. Then,*

$$\prod_{k=1}^n |t - t_k|^{\gamma_k} \in \mathscr{X}(\Gamma),$$

for all $\gamma_k > -\alpha_k$, $k = 1, 2, \ldots, n$.

Lemma 2.1.4. *Let $\mathscr{X}(\Gamma)$ be a BFS satisfying conditions (2.2) and (2.3) and Axioms 1 and 2. Then, the space $\mathscr{X}(\Gamma, \varrho)$, for $\varrho(t) = \prod_{k=1}^{n} |t - t_k|^{\gamma_k}$, $t_1, \ldots, t_k \in \Gamma$, satisfies conditions (2.2) and (2.3) as well if*

$$-\alpha(t_k) < \gamma_k < 1 - \beta(t_k), \quad k = 1, \ldots, n. \tag{2.22}$$

Let G be a piecewise continuous function on Γ $(G \in PC(\Gamma))$ such that $\inf_{t \in \Gamma} |G(t)| > 0$, and let $t_1, \ldots, t_n$ be the points of discontinuity of G. As usual, we put

$$\gamma(t) = \frac{1}{2\pi i} \ln \left(\frac{G(t-0)}{G(t+0)} \right) \tag{2.23}$$

and

$$\omega(t) = \prod_{k=1}^{n} (t - z_0)_k^{\gamma(t_k)}, \tag{2.24}$$

where $z_0 \in D^+$ and t_k are the discontinuity points of G and the functions $\omega_k(z) = (z - z_0)_k^{\gamma(t_k)}$ stand for univalent analytic functions in the complex plane with a cut passing from z_0 to infinity through the point $t_k \in \Gamma$. The function

$$G_1(t) = \frac{G(t)}{\omega(t)} \tag{2.25}$$

is continuous on Γ irrespective of the choice of

$$\alpha_k := \Re\gamma(t_k) = \frac{1}{2\pi i} \arg \left(\frac{G(t-0)}{G(t+0)} \right).$$

Consider now the function

$$\varpi(z) = \prod_{k=1}^{m} \varpi_k(z) \quad \text{with} \quad \varpi_k(z) = \begin{cases} (z - t_k)^{\gamma(t_k)}, & z \in D^+, \\ \dfrac{(z - t_k)^{\gamma(t_k)}}{(z - z_0)^{\gamma(t_k)}}, & z \in D^-, \end{cases}$$

where the branch for the function $\left(\frac{z - t_k}{z - z_0} \right)^{\gamma(t_k)}$ is chosen so that it tends to one as $z \to \infty$, $z \in D^-$, so that ϖ_k is analytic in $D^\pm$.

Assume that the curve Γ at the points t_k has at least one-sided tangents. For $\beta_k := \Im\gamma(t_k)$, with γ_k chosen as in (2.22), and taking into account Lemmas 2.1.3 and 2.1.4, from the equality

$$\varpi_k(z) = e^{\alpha_k \ln |z - t_k| - \beta_k \arg(z - t_k)} e^{i(\beta_k \ln |z - t_k| + \alpha_k \arg(z - t_k))},$$

we conclude that

$$\varpi(z), \frac{1}{\varpi(z)} \in \mathscr{E}^1(D^\pm). \tag{2.26}$$

Let $X(z) = \varpi(z)X_1(z)$, where

$$X_1(z) = \exp(K(\ln(G_1(t)))),$$

and let us introduce a new function,

$$\Psi_1(z) = \frac{\Psi(z)}{\varpi(z)}. \tag{2.27}$$

As long as $\Psi \in \mathscr{E}^1(D^{\pm})$ and (2.26) holds, we have, according to Smirnov's theorem, that $\Psi_1 \in \mathscr{E}^1(D^{\pm})$. Since the equality $\Psi_1 = K(\Psi_1^+ - \Psi_1^-)$ holds, $\Psi_1 \in \mathscr{E}(D)$.

Now, for the function G_1 given in (2.25) and Ψ_1 in (2.27), let us consider the problem

$$\Psi_1^+(t) = G_1(t)\Psi_1^-(t) + \varpi(t)g(t). \tag{2.28}$$

Having resolved (2.28), we find that all the possible solutions of problem (2.9) in the case of $\aleph = \operatorname{ind} G_1(t) \geq 0$ are given by

$$\Psi(z) = \varpi(z)X_1(z)K\left(\frac{g}{\varpi^+ X_1^+}\right)(z) + \varpi(z)X(z)\rho(z), \tag{2.29}$$

where ρ is an arbitrary polynomial of degree $\aleph$.

Note that from Lemma 2.1.4, the function Ψ in (2.29) is such that $\Psi^{\pm} \in \mathscr{X}(\Gamma)$. Therefore, (2.29) with an arbitrary polynomial ρ provides the solutions for (2.9) in $\mathscr{E}(\Gamma)$. The case of a negative index is considered in the standard way.

Thus, we arrive at the following result.

Theorem 2.1.5. *Let Γ be a Lyapunov curve and $\mathscr{X}(\Gamma)$ be a BFS satisfying (2.2)–(2.4), (2.6), and Axioms 1 and 2. Let $G \in PC(\Gamma)$ such that $\inf_{t \in \Gamma}|G(t)| > 0$, with points of discontinuity $t_1, \ldots, t_n$, and suppose that the curve Γ at the points t_k has at least one-sided tangents. Let $\aleph = \operatorname{ind} G_1$, where G_1 is given by (2.25). For $\gamma(t_k)$ given in (2.23) satisfying (2.22), the statements of Theorem 2.1.2 holds for problem (2.9) if $X(z)$ is replaced by $X_1(z)$ and formula (2.16) is replaced by formula (2.29).*

2.1.3 $L^{\infty}(\Gamma)$-factorable coefficients

Now, we consider the BVP (2.10) with essentially measurable coefficients admitting a factorisation in the space $\mathscr{X}(\Gamma)$.

Theorem 2.1.6. *Let Γ be a Lyapunov curve, and let $\mathscr{X}(\Gamma)$ be a BFS satisfying (2.2)–(2.6). Let us suppose that the function G admits a factorisation*

$G(t) = G_-(t)t^\aleph G_+(t)$ *in* $\mathscr{X}(\Gamma)$. *Then:*

(a) *the boundary problem* (2.10) *is solvable if and only if*

$$G_-^{-1}g \in \mathscr{L}^1(\Gamma), \quad \varphi_0^- = G_-P_-G_-^{-1}g \in \overset{\circ}{\mathscr{X}}{}_-^1(\Gamma),$$
$$\varphi_0^+ = G_+^{-1}t^{-\aleph}P_+G_-^{-1}g \in \mathscr{X}_+(\Gamma); \tag{2.30}$$

(b) *if conditions* (2.30) *are fulfilled, then the general solution of problem* (2.10) *is of the form*

$$\varphi^+ = \varphi_0^+ + G_+^{-1}t^{-\aleph}\rho, \quad \varphi^- = \varphi_0^- - G_-\rho, \tag{2.31}$$

where ρ *is a polynomial of degree less than or equal to* $-\aleph - 1$ *if* $\aleph < 0$ *and equal to zero if* $\aleph \geq 0$.

Proof. Let us assume that $\{\varphi^+, \varphi^-\}$ is a solution of problem (2.10). Substituting the representation (2.8) of the function G into the boundary condition (2.10), we obtain

$$\varphi^-(t) + G_-(t)t^\aleph G_+(t)\varphi^+(t) = g(t)$$
$$f^-(t) + t^\aleph f^+(t) = G_-^{-1}(t)g(t), \tag{2.32}$$

where $f^- = G_-^{-1}\varphi^-$, $f^+ = G_+\varphi^+$. Since $G_-^{-1} \in \mathscr{X}_-'(\Gamma)$, $\varphi^- \in \overset{\circ}{\mathscr{X}}{}_-(\Gamma)$, we have, from the Hölder inequality (2.1), that $f^- \in \overset{\circ}{L}{}_-^1(\Gamma)$. Analogously, $f^+ \in L_+^1(\Gamma)$. Then, equation (2.32) implies $G_-^{-1}g \in \mathscr{L}^1(\Gamma)$; therefore, $P_+G_-^{-1}g \in L_+^1(\Gamma)$ and $P_-G_-^{-1}g \in \overset{\circ}{L}{}_-^1(\Gamma)$ are well defined.

Rewriting equation (2.32), we get

$$f^-(t) - P_-G_-^{-1}(t)g(t) = -t^\aleph f^+(t) + P_+G_-^{-1}(t)g(t).$$

Since f^- and $P_-G_-^{-1}g$ vanish at infinity, $L_+^1(\Gamma) \cap L_-^1(\Gamma) = $ Const. and $L_+^1(\Gamma) \cap \overset{\circ}{L}{}_-^1(\Gamma) = \{0\}$, then $\rho(t) = t^\aleph f^+(t) - P_+G_-^{-1}(t)g(t)$ is identically zero for $\aleph \geq 0$ and a polynomial of order less than or equal to $-\aleph - 1$ for $\aleph < 0$.

Put

$$f^-(t) = P_-G_-^{-1}(t)g(t) - \rho(t), \quad f^+(t) = t^{-\aleph}P_+G_-^{-1}(t)g(t) + t^{-\aleph}\rho(t).$$

Returning to the functions $\varphi^\pm$, we have formulas (2.31). According to the assumptions, $\{\varphi^+, \varphi^-\}$ is a solution to problem (2.10). Thus, in the boundary Γ, $\varphi^+ \in \mathscr{X}_+(\Gamma)$ and $\varphi^- \in \overset{\circ}{\mathscr{X}}_-(\Gamma)$. Since $\rho \in \overset{\circ}{L^\infty_-}(\Gamma)$ and $t^{-\aleph}\rho \in L^\infty_+(\Gamma)$, the conditions $G_-\rho \in \overset{\circ}{\mathscr{X}}_-(\Gamma)$ and $G_+^{-1}t^{-\aleph}\rho \in \mathscr{X}_+(\Gamma)$ are satisfied; therefore,

$$\varphi_0^+ \in \mathscr{X}_+(\Gamma), \quad \varphi_0^- \in \overset{\circ}{\mathscr{X}}_-(\Gamma),$$

which proves the necessity of conditions (2.30) for the solvability of problem (2.10) as well as the fact that every solution of the problem is of the form given by (2.31).

Proving that conditions (2.30) imply that every pair $\{\varphi^+, \varphi^-\}$ defined by (2.31) is a solution of problem (2.10) can be directly carried out by evaluating functions (2.31) into (2.10), taking into consideration the boundedness of the functions on the corresponding spaces, which is given by conditions (2.30).   □

From Cauchy's formula (Golozin, 1969, Chapter X, Section 4, Theorem 1), for any analytic function representable by an integral of Cauchy type on a domain D, we immediately obtain the following result.

Corollary 2.1.7. *For the solvability of problem* (2.10)*, it is necessary that, for* $\aleph \geq 0$*, the conditions*

$$\int_\Gamma G_-^{-1}(\tau)g(\tau)\tau^{-\kappa}d\tau = 0, \quad \kappa = 1, \ldots, \aleph,$$

are valid.

Remark 2.1.8.

(1) Note that to establish Theorem 2.1.6, it was only necessary to validate condition (a) in the notion of factorisation of a essentially bounded function in $\mathscr{X}(\Gamma)$. Thus, in this sense, the factorisation used in Theorem 2.1.6 is weaker than that defined on p. 15.

(2) Similar conclusions pertaining to those given in Theorem 2.1.6 can be drawn for problem (2.9) by considering the associate space $\mathscr{X}'(\Gamma)$ of $\mathscr{X}(\Gamma)$. In this case, we define the factorisation in $\mathscr{X}(\Gamma)$ for $a \in \mathcal{G}(L^\infty(\Gamma))$ as

$$a(t) = a_+(t)t^\aleph a_-(t)$$

$\aleph \in \mathbb{N}$, with $a_+ \in \mathscr{X}_+(\Gamma)$, $a_+^{-1} \in \mathscr{X}'_+(\Gamma)$, $a_- \in \mathscr{X}'_-(\Gamma)$, and $a_-^{-1} \in \mathscr{X}_-(\Gamma)$. The proof for this situation runs in a similar way to that of Theorem 2.1.6 with obvious changes.

2.2   Solvability Theory of SIEs with Factorable $L^\infty(\Gamma)$ Coefficients

Now, we give the conditions that guarantee the existence of solutions for a class of SIEs with essentially bounded functions as coefficients, admitting a factorisation in $\mathscr{X}(\Gamma)$ as (2.8). To attain our goals, the Fredholmness of the associated singular integral operator is studied.

2.2.1   *Simonenko's criterion for the Fredholm property of SIOs*

In this section, we present a Fredholm criterion for the operator $\mathcal{A} = aP_+ + bP_-$ on $\mathscr{X}(\Gamma)$, with $a, b \in \mathcal{G}(L^\infty(\Gamma))$. This is established by adapting the classical Simonenko's scheme for singular integral operators with generalised factorable functions on $L^p(\Gamma)$.

First, for a linear and bounded operator $\mathbb{A} \in \mathcal{B}(X, Y)$, the set $\ker \mathbb{A}$ of all solutions of the homogeneous equation

$$\mathbb{A}x = 0 \tag{2.33}$$

is referred to as the kernel of the operator $\mathbb{A}$. The dimension of the subspace $\ker \mathbb{A}$, i.e. the number of linearly independents solutions of equation (2.33) (called nullity), is denoted by $\alpha(\mathbb{A})$, and we write

$$\alpha(\mathbb{A}) := \dim \ker \mathbb{A}.$$

A bounded operator $\mathbb{A}$ is called normally solvable (in the sense of Hausdorff) if the equation

$$\mathbb{A}x = y$$

is solvable only for those elements y which are orthogonal to the equation $\mathbb{A}^* u = 0$, where $\mathbb{A}^*$ is the conjugate operator $\mathbb{A}^* : Y^* \longrightarrow X^*$ defined by the relation

$$(\mathbb{A}^* u)x = u(\mathbb{A}x).$$

That means $\mathbb{A}^* u = 0$ if and only if $u(y) = 0$, for all functions $u \in \ker \mathbb{A}^*$. This is equivalent to stating that the set image (range) of $\mathbb{A}$, denoted by $\operatorname{Im} \mathbb{A}$ and defined as $\operatorname{Im} \mathbb{A} = \{\mathbb{A}x : x \in X\}$, is a closed set.

For a normally solvable operator $\mathbb{A}$, the co-kernel of $\mathbb{A}$, $\operatorname{Coker} \mathbb{A}$, is defined as the quotient

$$\operatorname{Coker} \mathbb{A} = Y / \operatorname{Im} \mathbb{A}.$$

The dimension of this subspace (called deficiency) is denoted by

$$\beta(\mathbb{A}) := \dim \operatorname{Coker} \mathbb{A}.$$

The integer numbers $\alpha(\mathbb{A})$ and $\beta(\mathbb{A})$ are frequently called the deficiency numbers of $\mathbb{A}$.

An operator $\mathbb{A}$ is called a Fredholm operator, or Φ-operator, if $\alpha(\mathbb{A})$ and $\beta(\mathbb{A})$ are finite. In this case, the Fredholm index is defined by the number

$$\operatorname{Ind} \mathbb{A} := \alpha(\mathbb{A}) - \beta(\mathbb{A}).$$

The operator $\mathbb{A}$ is called semi-Fredholm if at least one of the numbers $\alpha(\mathbb{A})$ or $\beta(\mathbb{A})$ is finite.

For the sake of completeness, we present the following classical result, which will be useful for proving some results.

Theorem 2.2.1 (Atkinson's Theorem). *Let X be a Banach space, and let A and B be Φ-operators. Then, BA is a Φ-operator and*

$$\operatorname{Ind} BA = \operatorname{Ind} B + \operatorname{Ind} A.$$

Let us consider first the singular integral operator $\mathcal{A}_a := aP_+ + P_-$ on $\mathscr{X}(\Gamma)$, with the coefficient $a \in L^\infty(\Gamma)$.

Proposition 2.2.2. *Suppose that $a \in L^\infty(\Gamma)$. If the operator $\mathcal{A}_a$ is semi-Fredholm, then $a \in \mathcal{G}(L^\infty(\Gamma))$.*

Proposition 2.2.3. *If $a \in \mathcal{G}(L^\infty(\Gamma))$, then $\min(\alpha(\mathcal{A}_a), \beta(\mathcal{A}_a)) = 0$.*

Propositions 2.2.2 and 2.2.3 can be proved as classical cases of $L^p(\Gamma)$ with minor modifications by using the well-known Lusin–Privalov theorem, which can be applied in this framework owing to assumption (2.2) and Lemma 2.0.3.

We can characterise the Fredholmness of the BVP (2.14) with an invertible continuous coefficient ba^{-1} through the Fredholmness of the operator $\mathcal{A} = aP_+ + bP_-$, given in the following criterion (Kokilashvili and Samko, 2003a, Theorem B).

Theorem 2.2.4. *Let $\mathscr{X}(\Gamma)$ be any BFS satisfying assumptions (2.2)–(2.4) and (2.6). The operator $\mathcal{A} = aP_+ + bP_-$ with $a, b \in C(\Gamma)$ is Fredholm in the space $\mathscr{X}(\Gamma)$ if and only if $a(t) \neq 0$ and $b(t) \neq 0$ for all $t \in \Gamma$. In this case, $\operatorname{Ind} \mathcal{A} = \operatorname{ind} \frac{b}{a} = \aleph$.*

Proof.

Step 1 (compactness of the commutators $aS - SaI$, $a \in C(\Gamma)$).
These commutators are compact in $\mathscr{X}(\Gamma)$. Indeed, it is known that
any function $a(t)$ continuous on Γ may be approximated in $C(\Gamma)$ by a
rational function $r(t)$, regardless of the Jordan curve Γ we are dealing
with, as is known from Mergelyan's result; see, for instance, Gaier
(1980, p. 169). Therefore, since the singular operator S is bounded in
$\mathscr{X}(\Gamma)$ by assumption (2.4), we obtain that the commutator $aS - SaI$ is
approximated to the operator norm in X by the commutator $rS - SrI$,
which is a finite-dimensional operator and, consequently, compact in
$\mathscr{X}(\Gamma)$. Therefore, $aS - SaI$ is compact.

Step 2 (sufficiency). By compactness of the commutators, we have
$(aP_+ + bP_-)(bP_+ + aP_-) = abI + T$, where T is a compact operator,
so the operator $(aP_+ + bP_-)$ has a regulariser. Consequently, it is
Fredholm.

Step 3 (the operator $\mathcal{A}_\aleph = P_+ + t^\aleph P_-$). Let $0 \in D^+$. The operator
$A_\aleph$ is right invertible in $\mathscr{X}(\Gamma)$ if $\aleph \geq 0$ and left invertible if $\aleph \leq 0$
and has the deficiency numbers $\alpha(A_\aleph) = \aleph$ and $\beta(A_\aleph) = 0$ if $\aleph \geq 0$
and $\alpha(A_\aleph) = 0$ and $\beta(A_\aleph) = |\aleph|$ for $k \leq 0$. Indeed, the operator
$\mathcal{A}_\aleph$ is Fredholm in $\mathscr{X}(\Gamma)$ by sufficiency (the previous step). The one-
sided invertibility follows from the relations $A_\aleph A_{-\aleph} = I$ if $\aleph \geq 0$ or
$A_{-\aleph} A_\aleph = I$ if $\aleph \leq 0$ well known on certain spaces and valid on $\mathscr{X}(\Gamma)$
by (2.4)–(2.6). To obtain the information on the deficiency numbers in
the space $\mathscr{X}(\Gamma)$, we observe that $H^\lambda(\Gamma) \subset C(\Gamma) \subset \mathscr{X}(\Gamma)$ by (2.2) and
that $\alpha(A_\aleph) = \aleph$ in the case of $\aleph \geq 0$ (Kokilashvili and Samko, 2002).
Therefore, $\alpha(A_\aleph) \geq \aleph$. Since $\mathscr{X}(\Gamma) \subset L^1(\Gamma)$, we also have $\alpha(A_\aleph) \leq \aleph$.
The case of $\aleph \leq 0$ is treated similarly.

Step 4 (the operator $\mathcal{N} = (t - \lambda)P_+ + P_-$). The operator $\mathcal{N}$ is
invertible in $\mathscr{X}(\Gamma)$ if $\lambda \in D^-$ and is Fredholm with $\operatorname{Ind}\mathcal{N} = -1$ if
$\lambda \in D^+$. Indeed, the invertibility in the case when $\lambda \in D^-$ is checked
directly: $\mathcal{N}_1 \mathcal{N} = \mathcal{N}\mathcal{N}_1 = I$, where $\mathcal{N}_1 = \frac{1}{t-\lambda}P_+ + P_-$, with conditions
(2.4)–(2.6) taken into account. The case when $\lambda \in D^+$ follows from
step 3 since $(t - \lambda)P_+ + P_- = (t - \lambda)[P_+ + (t - \lambda)^{-1}P_-]$.

Step 5 (necessity). Suppose that $a(t_0) = 0$ for some $t_0 \in \Gamma$ and the
operator A is Fredholm. By the compactness of the commutators $aS -$
SaI (step 1), we have the relations $aP_+ + bP_- = (P_+ + bP_-)(aP_+ + P_-) +$
$T_1 = (aP_+ + P_-)(P_+ + bP_-) + T_2$, where T_1 and T_2 are compact operators
in $\mathscr{X}(\Gamma)$. So, $aP_+ + P_-$ is Fredholm and $a(t_0) = 0$. We may approximate
the function a in $C(\Gamma)$ by the rational functions a_ε such that $a_\varepsilon(t_0) = 0$.
Then, the operators $a_\varepsilon P_+ + P_-$ with ε small enough are Fredholm.

To arrive at a contradiction, we follow Gohberg and Krupnik (1992b, p. 174) and represent a_ε as $a_\varepsilon(t) = (t - t_0)s(t)$. Then, $a_\varepsilon P_+ + P_- = (sP_+ + P_-)[(t - t_0)P_+ + P_-] = [(t - t_0)P_+ + P_-](sP_+ + P_-) + T$, where T is a compact operator. Therefore, the operator $(t - t_0)P_+ + P_-$ has a regulariser and is a Fredholm operator, which is impossible in view of the statement in step 4 and the known property of the stability of index of the Fredholm operator.

Step 6 (index formula). As done by Gohberg and Krupnik (1992b, p. 103), we approximate the function $c(t) = \frac{a(t)}{b(t)}$ by a rational function $r(t)$ so that

$$c(t) = r(t)[1 + m(t)] \quad \text{with} \ \max_{t \in \Gamma} |m(t)| < \frac{1}{||P_+||}. \tag{2.34}$$

Let $r(t) = t^{-\aleph} \frac{\chi_+(t)}{\chi_1(t)}$ be the factorisation of the function $r(t)$. Since $\|m\| < 1$, we have $\mathrm{ind}(1 + m) = 0$, and then $\mathrm{ind}\, r = \mathrm{ind}\, c = -\aleph$. In the case of $\aleph \le 0$, the representation is valid:

$$\mathcal{A} = b\chi_-(I + mP_+)\left(\frac{1}{\chi_+}P_+ + \frac{1}{\chi_-}P_-\right)(t^{-\aleph}P_+ + P_-), \tag{2.35}$$

with reference to conditions (2.4)–(2.6). The operator $I + mP_-$ is invertible since $\|mP_+\| < 1$ by (2.34) and (2.4). Since the operator is obviously invertible in $\mathscr{X}(\Gamma)$, from (2.35) we obtain $\mathrm{Ind}\,\mathcal{A} = \mathrm{ind}(t^{\aleph}P_+ + P_-) = \aleph$ according to the statement in step 3. $\qquad\square$

The following is a Fredholm criterion for the operator $\mathcal{A}_a$ on the space $\mathscr{X}(\Gamma)$. This result was established on the spaces $L^p(\Gamma)$ by Simonenko (1964, 1968; see also, for example, Mikhlin and Prössdorf, 1986) and by Karlovich (1998) in the case of reflexive rearrangement-invariant spaces. The proof that we give here follows analogously to these cases.

Theorem 2.2.5. *Let Γ be a Lyapunov curve. For a function $a \in L^\infty(\Gamma)$ to admit a factorisation in $\mathscr{X}(\Gamma)$, it is necessary and sufficient that the operator $\mathcal{A}_a = aP_+ + P_-$ be a Φ-operator on $\mathscr{X}(\Gamma)$. In that case, $\mathrm{Ind}\,\mathcal{A}_a = -\,\mathrm{ind}\, a$.*

Proof.

Necessity. Let $0 \in D^+$. First, we assume that the function a admits a factorisation $a = a_- a_+$ in $\mathscr{X}(\Gamma)$. Let $r \in \mathcal{R}(\Gamma)$. From Lemma 2.0.4 and the definition of the factorisation, we have that $a_+^{-1}P_+ a_-^{-1}r \in \mathscr{X}_+(\Gamma)$ and $a_- P_- a_-^{-1}r \in \overset{\circ}{\mathscr{X}}_-(\Gamma)$. Consider the bounded linear operator

$$\mathcal{B} := (a_+^{-1}P_+ + a_- P_-)a_-^{-1}I = I + (1 - a)a_+^{-1}P_+ a_+ a^{-1}I.$$

Then,

$$\mathcal{A}_a \mathcal{B} r = (aP_+ + P_-)(a_+^{-1}P_+ + a_-P_-)a_-^{-1}Ir$$

$$= aa_+^{-1}P_+a_-^{-1}r + a_-P_-a_-^{-1}r = r.$$

Analogously, $\mathcal{B}\mathcal{A}_a r = r$. Since $\mathcal{B}$ is bounded in $\mathscr{X}(\Gamma)$, due to assumptions (2.3) and (2.4) and because $\mathcal{R}(\Gamma)$ is dense in $\mathscr{X}(\Gamma)$, we conclude that $\mathcal{A}_a$ is invertible with inverse $\mathcal{A}_a^{-1} = \mathcal{B}$. Hence, $\operatorname{Ind}\mathcal{A}_a = 0$.

Now, let $a(t) = a_-(t)t^\aleph a_+(t)$ be a factorisation in $\mathscr{X}(\Gamma)$. Then, the function $at^{-\aleph}$ admits the factorisation $at^{-\aleph} = a_-a_+$ in $\mathscr{X}(\Gamma)$. Thus, the operator $at^{-\aleph}P_+ + P_-$ is invertible in $\mathscr{X}(\Gamma)$. If $\aleph > 0$, then $\mathcal{A}_a$ can be represented in the form

$$\mathcal{A}_a = (at^{-\aleph}P_+ + P_-)(t^\aleph P_+ + P_-).$$

From step 3 in the proof of Theorem 2.2.4, we have that $t^\aleph P_+ + P_- = t^\aleph(P_+ + t^{-\aleph}P_-)$ is a Φ-operator with index $-\aleph$. Then, in virtue of the Atkinson theorem for Fredholm operators, we conclude that $\mathcal{A}_a$ is a Φ-operator with index $-\aleph$.

In the case of $\aleph < 0$, we have

$$at^{-\aleph}P_+ + P_- = \mathcal{A}_a(t^{-\aleph}P_- + P_-),$$

with $t^{-\aleph}P_+ + P_- = t^{-\aleph}(P_+ + t^\aleph P_-)$ being a Φ-operator with index $\aleph$, and thus $\mathcal{A}_a$ is a Φ-operator with index $-\aleph$ too.

Sufficiency. Suppose that the operator $\mathcal{A}_a$ is a Fredholm operator with index $-\aleph$. By Proposition 2.2.2, $a \in \mathcal{G}(L^\infty(\Gamma))$. Consider the operator $\mathcal{A}_q$, where $q(t) = a(t)t^{-\aleph}$. By the compactness of the commutator $aS + SaI$ for $a \in C(\Gamma)$ given by step 1 of the proof of Theorem 2.2.4, $\mathcal{A}_q = \mathcal{A}_a\mathcal{A}_{t^{-\aleph}} + \mathcal{K}$, where $\mathcal{K}$ is a compact operator. Hence, the operator $\mathcal{A}_q$ is Fredholm, and $\operatorname{Ind}\mathcal{A}_q = 0$. From the Atkinson theorem and the fact that the Fredholm index is invariant under compact perturbations, then, by Proposition 2.2.3, it follows that $\mathcal{A}_q$ is invertible in $\mathscr{X}(\Gamma)$. Taking into account the fact that the operator S^* adjoint to the operator S in the dual space $\mathscr{X}^*(\Gamma)$ is given by $S^* = -HSH$ and applying Lemma 2.0.3, we can show that in this case the operator $\mathcal{A}_{q^{-1}}$ is invertible in the associate space $\mathscr{X}'(\Gamma)$.

Let $\varphi_0 \in \mathscr{X}(\Gamma)$ and $\psi_0 \in \mathscr{X}'(\Gamma)$ be the solutions of the equations $\mathcal{A}_q\varphi = 1$ and $\mathcal{A}_{q^{-1}}\psi = 1$, respectively. Applying Lemma 2.0.4, one can show that $a_+ := P_+\psi_0$ and $a_- := 1 - P_-\psi_0$ are the factors of the factorisation $a(t) = a_-(t)t^\aleph a_+(t)$, that is, $a_- \in \mathscr{X}_-(\Gamma)$, $a_+ \in \mathscr{X}'_+(\Gamma)$, $a_-^{-1} \in \mathscr{X}'_-(\Gamma)$, and $a_+^{-1} \in \mathscr{X}_+(\Gamma)$. Now, we should prove the boundedness of $a_+^{-1}Sa_+I$

in $\mathscr{X}(\Gamma)$. Without loss of generality, assume that $\|q\|_\infty < 1$, and consider the operator

$$\mathcal{B} := (a_+^{-1}P_+ + a_-P_-)a_-^{-1}I = I + (1-q)a_+P_+a_-^{-1}I.$$

As above, $\mathcal{B} = \mathcal{A}_q^{-1}$ is bounded; therefore, $a_+^{-1}P_+a_-^{-1}I$ is bounded too, which is equivalent to $a_+^{-1}Sa_-^{-1}I$ being bounded in $\mathscr{X}(\Gamma)$. $\square$

2.2.2 Effective solutions of SIEs

The solvability theory in the space $\mathscr{X}(\Gamma)$ of the equation

$$\mathcal{A}\varphi(t) := u(t)\varphi(t) + \frac{v(t)}{\pi i}\text{p.v.}\int_\Gamma \frac{\varphi(\tau)}{\tau - t}d\tau = f(t), \quad t \in \Gamma, \quad u,v \in L^\infty(\Gamma),$$
(2.36)

or, alternatively,

$$\mathcal{A} = aP_+ + bP_-, \quad a := u + v, \ b := u - v,$$

is given in the following results.

Theorem 2.2.6. *Let Γ be a Lyapunov curve, $a,b \in L^\infty(\Gamma)$, and let $\mathscr{X}(\Gamma)$ be a BFS satisfying (2.2)–(2.6). Then, for the operator $\mathcal{A} = aP_+ + bP_-$ to be a Φ_+ or Φ_- operator on $\mathscr{X}(\Gamma)$, it is necessary that $a,b \in \mathcal{G}(L^\infty(\Gamma))$. Let a and b be invertible functions on $L^\infty(\Gamma)$. Then, $\mathcal{A}$ is a Φ operator if and only if the function ab^{-1} admits a factorisation (2.8). Let $\mathcal{A}$ be a Φ operator and $\aleph = \text{ind}\,\frac{a}{b}$. Then, $\text{Ind}\,\mathcal{A} = -\aleph$, and the operator $\mathcal{A}$ is left-sided invertible, right-sided invertible, or two-sided invertible if $\aleph > 0$, $\aleph < 0$, or $\aleph = 0$, respectively. The corresponding (one- or two-sided) inverse is of the form*

$$\mathcal{A}^{-1} = (t^{-\aleph}P_+ + P_-)(c_+^{-1}P_+ + c_-P_-)c_-^{-1}b^{-1}I,$$

where $ab^{-1} = c_-t^{\aleph}c_+$ is the factorisation in $\mathscr{X}(\Gamma)$ of the function ab^{-1}.

Proof. Let $0 \in D^+$, and let us assume that $a,b \in \mathcal{G}(L^\infty(\Gamma))$. Then, the operator $\mathcal{A}$ can be written as

$$\mathcal{A} = b(ab^{-1}P_+ + P_-),$$

where the multiplicator operator bI is bounded in $\mathscr{X}(\Gamma)$ by assumption (2.3) and invertible with inverse $b^{-1}I$. Thus, it is a Φ operator.

Since $ab^{-1} \in \mathcal{G}(L^\infty(\Gamma))$, then from Proposition 2.2.2, the operator $ab^{-1}P_+ + P_-$ is semi-Fredholm. Therefore, by the Atkinson theorem, $\mathcal{A}$ is a semi-Fredholm operator.

On the other hand, Theorem 2.2.5 guarantees that $ab^{-1}P_+ + P_-$ is a Φ operator if and only if ab^{-1} admits a factorisation. Then, reasoning as before, we conclude that $\mathcal{A} = b(ab^{-1}P_+ + P_-)$ is a Φ operator iff ab^{-1} admits a factorisation in $\mathscr{X}(\Gamma)$.

Now, let $\mathcal{A}$ be a Φ operator and $\aleph = \operatorname{ind} ab^{-1}$. Since bI is invertible, $\operatorname{Ind} bI = 0$, and from the Atkinson theorem,

$$\operatorname{Ind} \mathcal{A} = \operatorname{Ind} bI + \operatorname{Ind}(ab^{-1}P_+ + P_-) = \operatorname{Ind}(ab^{-1}P_+ + P_-).$$

Theorem 2.2.5 asserts that $\operatorname{Ind}(ab^{-1}P_+ + P_-) = -\operatorname{ind} ab^{-1} = -\aleph$.

To prove the (one- or two-sided) invertibility of $\mathcal{A}$, note that for $ab^{-1} = c_- t^{\aleph} c_+$,

$$\mathcal{A} = bc_-(t^{\aleph}c_+P_+ + c_-^{-1}P_-) = bc_-(t^{\aleph}P_+ + P_-)(c_+P_+ + c_-^{-1}P_-), \quad (2.37)$$

with bc_-I and $c_+P_+ + c_-^{-1}P_-$ being invertible operators with inverses given by

$$(bc_-I)^{-1} = b^{-1}c_-^{-1}I, \quad (c_+P_+ + c_-^{-1}P_-)^{-1} = c_+^{-1}P_+ + c_-P_-.$$

From step 3 in the proof of Theorem 2.2.4, the operator $t^{\aleph}P_+ + P_-$ is left-, right-, or two-sided invertible (and so, by equality (2.37), the operator $\mathcal{A}$) if $\aleph > 0$, $\aleph < 0$, or $\aleph = 0$, respectively. Direct computations show that the form of the inverses of $\mathcal{A}$ are in fact given by $(t^{-\aleph}P_+ + P_-)(c_+^{-1}P_+ + c_-P_-)c_-^{-1}b^{-1}I$. $\square$

The following result gives the dimension of $\ker \mathcal{A}$ and $\operatorname{Coker} \mathcal{A}$, as well as the solvability conditions for equation (2.36).

Theorem 2.2.7. *Let Γ be a Lyapunov curve, let $\mathscr{X}(\Gamma)$ be a BFS satisfying (2.2)–(2.6), and let $a, b \in \mathcal{G}(L^{\infty}(\Gamma))$. Moreover, assume that the function ab^{-1} admits a factorisation $ab^{-1} =: c = c_- t^{\aleph} c_+$ in the space $\mathscr{X}(\Gamma)$. Then, if $\aleph = \operatorname{ind} c < 0$,*

$$\ker(aP_+ + bP_-) = \operatorname{span}\{g, gt, \ldots, gt^{|\aleph|-1}\}, \quad (2.38)$$

where $g = c_+^{-1} - c_- t^{\aleph}$. In the case of $\aleph > 0$,

$$\operatorname{Coker}(aP_+ + bP_-) = \operatorname{span}\{bc_-, bc_- t, \ldots, bc_- t^{\aleph-1}\}, \quad (2.39)$$

and the equation $aP_+\varphi + bP_-\varphi = f$ has a solution if and only if

$$\int_{\Gamma} f(t)b^{-1}(t)c_-^{-1}(t)t^{-j} dt = 0, \quad j = 1, 2, \ldots, \aleph. \quad (2.40)$$

Proof. Let $\mathcal{A} = aP_+ + bP_-$, and let us assume that $0 \in D^+$. From identity (2.37), we have that $\mathcal{A}$ and $t^\aleph P_+ + P_-$ are equivalent operators; therefore,

$$\dim \ker \mathcal{A} = \dim \ker(t^\aleph P_+ + P_-) = \dim \ker(P_+ + t^{-\aleph} P_-).$$

Step 3 in the proof of Theorem 2.2.4 gives us that if $\aleph \leq 0$, $\dim \ker \mathcal{A} = |\aleph|$ and if $\aleph \geq 0$, then $\dim \operatorname{Coker} \mathcal{A} = \aleph$.

Now, let us suppose that $\aleph < 0$. We now aim to find the set $\ker(P_+ + t^{-\aleph} P_-)$. That is,

$$\{\varphi \in \mathscr{X}(\Gamma) \ : \ P_+\varphi + t^{-\aleph} P_- \varphi = 0\}.$$

Since $\dim \ker(P_+ + t^{-\aleph} P_-) = |\aleph|$, there is a polynomial $p_{|\aleph|-1}(t) = a_{|\aleph|-1} t^{|\aleph|-1} + \cdots + a_1 t + a_0$ of degree at most $|\aleph| - 1$ such that $t^{|\aleph|} P_- \varphi + p_{|\aleph|-1} \in P_-(\mathscr{X}(\Gamma))$. From the above, we have that

$$P_+\varphi = p_{|\aleph|-1}, \quad P_-\varphi = -\frac{p_{|\aleph|-1}}{t^{|\aleph|}}.$$

Thus, we get

$$P_+\varphi(t) + P_-\varphi(t) = \varphi(t) = p_{|\aleph|-1}(t)\left[1 - \frac{1}{t^{|\aleph|}}\right];$$

therefore,

$$\ker(P_+ + t^{-\aleph} P_-) = \operatorname{span}\left\{t^{|\aleph|-1} - \frac{1}{t}, t^{|\aleph|-2} - \frac{1}{t^2}, \ldots, 1 - \frac{1}{t}\right\}.$$

On the other hand, by (2.37), we obtain

$$\ker \mathcal{A} = (c_+^{-1} P_+ + c_- P_-) \ker(t^\aleph P_+ + P_-) = \operatorname{span}\{g_1, \ldots, g_{|\aleph|}\},$$

with $g_j = (c_+^{-1} P_+ + c_- P_-)(t^{|\aleph|-j} - t^{-j})$, $j = 1, \ldots, |\aleph|$. Here,

$$\begin{aligned}
g_j &= (c_+^{-1} P_+ + c_- P_-)(t^{|\aleph|-j} - t^{-j}) \\
&= (c_+^{-1} P_+ + c_- P_-)t^{|\aleph|-j} - (c_+^{-1} P_+ + c_- P_-)t^{-j} \\
&= c_+^{-1} P_+ t^{|\aleph|-j} - c_- P_- t^{-j} \\
&= c_+^{-1} t^{|\aleph|-j} - c_- t^{-j}.
\end{aligned}$$

This proves equality (2.38).

Now, let us assume that $\aleph > 0$. From (2.37), we have that $\operatorname{Im} \mathcal{A} = bc_- \operatorname{Im}(T^\aleph P_+ + P_-)$, which gives us (2.39) because $\operatorname{Im}(T^\aleph P_+ + P_-)$ consists of all functions $\varphi \in \mathscr{X}(\Gamma)$ such that $P_+\varphi$ has a zero of order at most $\aleph$ at the point $t = 0$. On the other hand, from (2.37) and step 3 of Theorem 2.2.4, $\mathcal{A}$ is a Φ operator that is left invertible and therefore normally solvable, so the equation $\mathcal{A}\varphi = f$ has a solution if and only if

$$\int_\Gamma f(t)\overline{y_j(t)}|dt| = 0, \quad j = 1, \ldots, m, \tag{2.41}$$

where $y_1, \ldots, y_m$ are all the linearly independent solutions of the adjoint homogeneous equation $\mathcal{A}^*y = 0$ in $\mathscr{X}(\Gamma)$. From Lemma 2.0.3, the adjoint operator of $\mathcal{A}$ is defined by $\mathcal{A}^* = H(P_+b + P_-a)H$ because $P_+^* = HP_-H$ and $P_-^* = HP_+H$. Therefore,

$$H\mathcal{A}^*z_j = (P_+b + P_-a)c_-^{-1}b^{-1}t^{-j} = P_+c_-^{-1}t^{-j} + P_-c_+t^{\aleph-j} = 0,$$

so $z_j \in \ker \mathcal{A}^*$, $j = 1, \ldots, \aleph$. However, the functions z_j are linearly independent, and since

$$\dim \ker \mathcal{A}^* = \dim \operatorname{Coker} \mathcal{A} = \aleph,$$

we conclude that $\aleph = m$ and $z_j = y_j$, for all j.

Besides, $\overline{y_j(t)}|dt| = h(t)c_-^{-1}(t)b^{-1}(t)t^{-j}|dt| = c_-^{-1}(t)b^{-1}(t)t^{-j}dt$; consequently, (2.40) and (2.41) coincide, which completes the proof. $\square$

2.3 On the Fredholmness of BVPs and the Explicit Representation of the Solutions of the Corresponding SIEs with $L^\infty(\Gamma)$ Coefficients

In this section, we analyse the Fredholm property of BVPs with continuous, piecewise continuous, or factorable essentially bounded coefficients. Also, the explicit representation of the solutions of the corresponding SIEs are stated.

2.3.1 *Continuous coefficients*

As a consequence of Theorem 2.2.4, we have the following result.

Proposition 2.3.1. *Let Γ be a Lyapunov curve, and let $\mathscr{X}(\Gamma)$ be any BFS satisfying assumptions (2.2)–(2.4) and (2.6). The BVP (2.14) with a continuous coefficient ba^{-1} is Fredholm with index $\aleph$ if and only if $ba^{-1} \in \mathcal{G}(C(\Gamma))$ with $\operatorname{ind} ba^{-1} = \aleph$.*

On the other hand, using the equivalence mentioned above, we can give an explicit representation of the solutions of equation (2.36) with continuous coefficients a and b, satisfying the assumptions in Theorem 2.2.4, through the two-term BVP (2.14).

Theorem 2.3.2. *Let Γ be a Lyapunov curve, and let $\mathscr{X}(\Gamma)$ be any BFS satisfying assumptions (2.2)–(2.4) and (2.6). Equation (2.36) with continuous coefficients a and b has solutions if and only if $a(t) \neq 0$ and $b(t) \neq 0$, for all $t \in \Gamma$. The solutions are described as follows, depending on the different situations:*

(a) $(\aleph \geq 0)$

$$\varphi(t) = \frac{(1 - (t - z_0)^{\aleph})e^{h(t)}}{2\pi i} \int_\Gamma \frac{f(\tau)}{e^{h(\tau)}} \frac{d\tau}{\tau - z} + (1 - (t - z_0)^{\aleph})e^{h(t)}\rho(t),$$

(2.42)

where $h(t) = K\left(\ln \frac{b(\tau)}{a(\tau)}(\tau - z_0)^{\aleph}\right)(t)$ and ρ is an arbitrary polynomial of degree $\aleph - 1$.

(b) $(\aleph < 0)$ *The unique solution in this case is as in (2.42) with $\rho(t) \equiv 0$. In addition, it is necessary that*

$$\int_\Gamma \frac{f(\tau)\tau^\kappa}{e^{h(\tau)}} d\tau = 0, \quad \kappa = 0, \ldots, |\aleph| - 1.$$

(2.43)

Here, $\aleph := \operatorname{ind} ba^{-1}$.

Proof. From Theorem 2.2.4, the conditions $a(t) \neq 0$ and $b(t) \neq 0$, for all $t \in \Gamma$, are equivalent to the Fredholmness of the associated operator $\mathcal{A} = aP_+ + bP_-$. Therefore, equation (2.36) is solvable. From Theorem 2.1.2, the solutions of the boundary problem (2.14) are given by formulas (2.16). The solvability conditions (2.43) are necessary for the solvability of problem (2.14), as stated in Theorem 2.1.2. Finally, from the Sokhotski–Plemelj formulas, we have (2.42). □

2.3.2 *Piecewise continuous coefficients*

The Fredholmness of the boundary problem (2.14) with piecewise continuous coefficients is characterised by using the Fredholmness of the operator $\mathcal{A} = aP_+ + bP_-$ with coefficients in the same class. To establish this, we use the so-called Khvedelidze–Gohberg–Krupnik investigation scheme. We begin with this approach by recalling the reformulations of the notions of p-nonsingularity and p-index in the framework of BFSs introduced by Kokilashvili and Samko (2003a).

For a BFS $\mathscr{X}(\Gamma)$ satisfying Axiom 1, a function $G \in PC(\Gamma)$ is called $\mathscr{X}(\Gamma)$-*nonsingular* if $\inf_{t \in \Gamma} |G(t)| > 0$ and

$$\frac{1}{2\pi} \arg \frac{G(t_k - 0)}{G(t_k + 0)} \notin [\alpha(t_k), \beta(t_k)] + \mathbb{Z},$$

where $[\cdots] + \mathbb{Z}$ stands for the set $\cup_{\xi \in [\cdots]} \{\xi, \xi \pm 1, \xi \pm 2, \dots\}$ and α and β are the index functions of the space $\mathscr{X}(\Gamma)$. For a $\mathscr{X}(\Gamma)$-nonsingular function, the integer

$$\text{ind } a = \sum_{k=1}^{n} [\theta(t_k) - \Re\gamma(t_k)],$$

where $\theta(t_k)$ are the increments

$$\theta(t_k) = \frac{1}{2\pi} \int_{t_k + 0}^{t_{k+1} - 0} d \arg G(t),$$

will be referred to as $\mathscr{X}(\Gamma)$-*index* of the function G.

We recall that $\Re\gamma(t_k)$, and the arguments $\theta(t_k)$ are chosen in the interval

$$\beta(t_k) - 1 < \Re\gamma(t_k) < \alpha(t_k). \tag{2.44}$$

The Fredholm criterion of the operator $\mathcal{A} = aP_+ + bP_-$ reads as follows (Kokilashvili and Samko, 2003a, Thoerem C).

Theorem 2.3.3. *Let $\mathscr{X}(\Gamma)$ be any BFS satisfying (2.2)–(2.4), (2.6), and Axioms 1 and 2. The operator $\mathcal{A} = aP_+ + bP_-$ with $a, b \in PC(\Gamma)$ is Fredholm in the space $\mathscr{X}(\Gamma)$ if*

$$\inf_{t \in \Gamma} |a(t)| \neq 0, \quad \inf_{t \in \Gamma} |b(t)| \neq 0 \tag{2.45}$$

and the function

$$\frac{a}{b} \quad \text{is } \mathscr{X}(\Gamma)\text{-nonsingular.} \tag{2.46}$$

In this case,

$$\text{Ind } \mathcal{A} = -\text{ind } \frac{a}{b}. \tag{2.47}$$

Condition (2.45) is also necessary for the operator $\mathcal{A}$ to be Fredholm in $\mathscr{X}(\Gamma)$. If the index functions α and β of the space $\mathscr{X}(\Gamma)$ coincide at the points of discontinuity t_k of the coefficients a and b,

$$\alpha(t_k) = \beta(t_k), \quad k = 1, 2, \dots, n,$$

then condition (2.46) is necessary as well.

Proof. Because of condition (2.45), we may assume that $b(t) \equiv 1$ (where the necessity of (2.45) for both a and b is simultaneously shown similarly to the case of $b(t) \equiv 1$).

Sufficiency. Let

$$\omega(t) = \frac{\omega^+(t)}{\omega^-(t)} \quad \omega^+(t) = \prod_{k=1}^{n}(z - t_k)^{\gamma(t_k)}, \quad \omega^-(t) = \prod_{k=1}^{n}\left(\frac{z - t_k}{z - z_0}\right)^{\gamma(t_k)}$$

be the well-known factorisation of function (2.24). We recall that $\Re\gamma(t_k)$ are chosen according to (2.44). We make use of the well-known representation

$$aP_+ + P_- = \frac{1}{\omega^-}(a_1 P_+ + P_-)\omega^-(\omega P_+ + P_-), \qquad (2.48)$$

where a_1 is function (2.25); see, for instance, Karapetiants and Samko (2001, p. 22). The function a_1 is in $C(\Gamma)$ by the choice of the values $\gamma(t_k)$. Relation (2.48) being valid, for instance, in the case of 'nice' functions is extended to the space $\mathscr{X}(\Gamma)$ by condition (2.6) since both the operators $\omega P_+ + P_-$ and $\frac{1}{\omega}(a_1 P_+ + P_-)\omega^-$ are bounded in $\mathscr{X}(\Gamma)$: the former by condition (2.4), and the latter by Lemma 2.1.4. The operator $\frac{1}{\omega^-}(a_1 P_+ + P_-)\omega^-$ is Fredholm in $\mathscr{X}(\Gamma)$ by Theorem 2.2.4 and Lemma 2.1.4, and its index in $\mathscr{X}(\Gamma)$ is equal to $\mathrm{ind}a_1$, which is nothing but $\mathrm{ind}a$. Thus, (2.47) is obtained.

It remains to show that the operator $\omega P_+ + P_-$ is invertible in the space $\mathscr{X}(\Gamma)$ thanks to the choice (2.44). This is checked as follows: $N(\omega P_+ + P_-) = (\omega P_+ + P_-)N$, where $N = \frac{1}{\omega^-}\left(\frac{1}{\omega}P_+ + P_-\right)\omega^-$. The operator K is bounded under the choice of (2.44) in the space $\mathscr{X}(\Gamma)$ by Lemma 2.1.4.

Necessity. Let the operator $\mathcal{A}$ be Fredholm in $\mathscr{X}(\Gamma)$. We first assume that $a(t_k \pm 0) \neq 0$, $k = 1, 2, \ldots, n$. We need to show that $a(t) \neq 0$ for all other points and that the required conditions on the jumps are satisfied:

Step 1 (reduction to a simpler operator). Since $a(t_k \pm 0) \neq 0$, the function $\omega(t)$ is well defined and the function $a_1(t) = \frac{a(t)}{\omega(t)}$ is continuous. As the commutators $aS - SaI$ and $a \in C(\Gamma)$ are compact in the space $\mathscr{X}(\Gamma)$ (see step 1 in the proof of Theorem 2.2.4), we have

$$\mathcal{A} = (\omega P_+ + P_-)(a_1 P_+ + P_-) + T. \qquad (2.49)$$

From the Fredholmness of the operator $\mathcal{A}$, we conclude by the Yood theorem (see, for example, Karapetiants and Samko, 2001, p. 4, Property 1.11) that the operator $\omega P_+ + P_-$ is a Φ_- operator.

Step 2 (the necessity of the conditions on jumps for the operator $\omega P_+ + P_-$). The following lemma reformulates a statement well known for $L^p(\Omega, \rho)$ spaces for the case of the abstract spaces $\mathscr{X}(\Gamma)$.

Lemma 2.3.4. *Let $a(t_k \pm 0) \neq 0$, $k = 1, 2, \ldots, n$, and the space $\mathscr{X}(\Gamma)$ satisfy conditions (2.2)–(2.6) and Axioms 1 and 2, and let $\alpha(t_k) = \beta(t_k)$, $k = 1, 2, \ldots, n$. The operator $\Psi = \omega P_+ + P_-$, with ω defined in (2.24), is a Φ_+ operator in the space $\mathscr{X}(\Gamma)$ if and only if*

$$\Re\gamma_k = \alpha(t_k) \,(\mathrm{mod}\,1), \quad \text{for all} \quad k = 1, 2, \ldots, n. \qquad (2.50)$$

Proof. By the sufficiency part of Theorem 2.3.3, condition (2.50) is sufficient. To prove the necessity, suppose that $\Re\gamma_k \alpha(t_k) + r$, for some $r = 0, \pm 1, \pm 2, \ldots$ and for some k, say $k = 1$, but that the operator Ψ is a Φ_+ or Φ_- operator. Let first $\Re\gamma_k \neq \alpha_k \,(\mathrm{mod}\,1)$ for all other $k = 2, 3, \ldots, n$. We put $\Psi_\varepsilon = \omega_{\pm\varepsilon} P_+ + P_-$, $\varepsilon > 0$, where $\omega_{\pm\varepsilon}(t) = (t - z_0)_1^{\pm\varepsilon} \omega(t)$. This new function has the new exponents $\gamma_1^{\pm\varepsilon} = \gamma_1 \pm \varepsilon$. We choose ε small enough so that $\Re\gamma_1 \pm \varepsilon - \alpha_1$ is not an integer. Then, by the sufficiency part of Theorem 2.3.3, the operators Ψ_ε and Ψ_ε are Fredholm operators in the space $\mathscr{X}(\Gamma, \rho)$. The calculation of the index using formula (2.47) gives

$$\mathrm{Ind}\,((t - z_0)^\nu P_+ + P_-) = [\alpha(t_1) - \Re\nu] \quad \text{in the case of} \quad \Re\nu \neq \alpha(t_1) + m,$$

where $m = 0, \pm 1, \pm 2 \ldots$ and $[\cdots]$ stands for the entire part of a number. Then,

$$\mathrm{Ind}\,\Psi_\varepsilon - \mathrm{Ind}\,\Psi_{-\varepsilon}$$
$$= [\Re\gamma(t_1) + \varepsilon - \alpha(t_1)] - [\Re\gamma(t_1) - \varepsilon - \alpha(t_1)] = [\varepsilon] - [-\varepsilon] = 1. \qquad (2.51)$$

However, $\|\Psi_{\pm\varepsilon} - \Psi\| \leq c \sup_{t \in \Gamma} |(t - z_0)^{\pm\varepsilon} - 1| \leq c_1\varepsilon$, which contradicts (2.51) by the stability theorem for Φ_+ operators in Banach spaces. This proves the lemma for the case of $k = 1$. If the condition in (2.50) is violated for several $k = n_1, \ldots, n_m$, the arguments are similar: the operators $\Psi_{\pm\varepsilon}$ must then be introduced with the functions $\omega_\pm(t) = \prod_{i=1}^m (t - z_0)_i^{\pm\varepsilon} \omega(t)$. $\qquad \square$

Step 3 (the necessity of the conditions for the operator N). Since the operator $P_+ + \omega P_-$ is a Φ_- operator (see step 1) by Lemma 2.3.4, conditions (2.50) are satisfied. Consequently, by the sufficiency part, the operator $P_+ + \omega P_-$ is a Fredholm operator in the space $\mathscr{X}(\Gamma)$. As is well known, if any two of the linear operators A, B, and AB are Fredholm, then the remaining one is Fredholm as well (see, for example, Karapetiants and Samko, 2001, p. 4, Property 1.12). Therefore, from (2.49), we conclude that the operator $a_1 P_+ + P_-$ is Fredholm in $\mathscr{X}(\Gamma)$. Then, by Theorem 2.2.4, $a_1(t) \neq 0$ and, consequently, $a(t) \neq 0$, $t \in \Gamma$.

Step 4. It remains to lift the assumptions $a(t_k \pm 0) \neq 0$ and $b(t_k \pm 0) \neq 0$. Suppose that some of the numbers $a(t_k \pm 0)$ are equal to zero and the operator $\mathcal{A}$ is Fredholm in $\mathscr{X}(\Gamma)$. There exists a complex number ε with an arbitrarily small modulus and a point t_0 close to t_k such that $a(t_k \pm 0) + \varepsilon \neq 0$ but $a(t_0) + \varepsilon = 0$. Let $\mathcal{A}_\varepsilon = (a + \varepsilon)P_+ + P_-$. Evidently, $\|\mathcal{A}_\varepsilon - \mathcal{A}\| = \|\varepsilon I\| = \varepsilon$. Therefore, by the stability theorem for Fredholm operators, we obtain that the operator $\mathcal{A}_\varepsilon$ is Fredholm for sufficiently small ε. This contradicts the preceding part. $\qquad\square$

The Fredholm property of the boundary problem (2.14) with a piecewise continuous coefficient is established in the following result.

Proposition 2.3.5. *Let Γ be a Lyapunov curve, and let $\mathscr{X}(\Gamma)$ be any BFS satisfying (2.2)–(2.4), (2.6) and Axioms 1 and 2. The BVP (2.14) with a piecewise continuous coefficient ab^{-1} is Fredholm with index $\aleph$ if*

$$\inf_{t \in \Gamma} |a(t)| \neq 0, \quad \inf_{t \in \Gamma} |b(t)| \neq 0, \tag{2.52}$$

and the function

$$\frac{a}{b} \quad \text{is } \mathscr{X}(\Gamma)\text{-nonsingular} \tag{2.53}$$

with index $\aleph = -\operatorname{ind} \frac{a}{b}$. Condition (2.52) is also necessary for the Fredholmness of problem (2.14). If the index functions α and β of the space $\mathscr{X}(\Gamma)$ coincide at the points of discontinuity t_k of the coefficients a and b,

$$\alpha(t_k) = \beta(t_k), \quad k = 1, 2, \ldots, n,$$

then condition (2.53) is necessary as well.

The representation of the solutions of equation (2.36) with the piecewise continuous coefficients a and b, satisfying the assumptions of Theorem 2.3.3, is given in the following result.

Theorem 2.3.6. *Let Γ be a Lyapunov curve, and let $\mathscr{X}(\Gamma)$ be any BFS satisfying (2.2)–(2.4), (2.6), and Axioms 1 and 2. Equation (2.36) with the* *piecewise continuous coefficients a and b has solutions if a and b satisfy (2.45) and ab^{-1} is $\mathscr{X}(\Gamma)$-nonsingular with discontinuity points $t_1, \ldots, t_n$; moreover, we suppose that the curve Γ at the points t_k has at least one-sided tangents. The solutions are described as follows, depending on the different situations:*

(a) ($\aleph \geq 0$)

$$\varphi(t) = (\omega^+(t)X_1^+(t) - \omega^-(t)X_1^-(t))K\left(\frac{f}{\omega^+X_1^+}\right)(t)$$

$$+ (\omega^+(t)X_1^+(t) - \omega^-(t)X_1^-(t))\rho(t), \qquad (2.54)$$

where

$$\omega^+(t) = \prod_{k=1}^{n}(t - t_k)^{\gamma(t_k)}, \quad \omega^-(t) = \prod_{k=1}^{n}\left(\frac{t - t_k}{t - z_0}\right)^{\gamma(t_k)},$$

$$X_1(t) = e^{K(\ln G_1(t))}, \quad G_1(t) = \frac{a(t)b^{-1}(t)}{\prod_{k=1}^{n}(t - t_k)^{\gamma(t_k)}},$$

with $z_0 \in D^+$ and ρ being an arbitrary polynomial of degree $\aleph - 1$.

(b) ($\aleph < 0$) *The unique solution in this case is as in (2.54) with $\rho(t) \equiv 0$. In addition, it is necessary that*

$$\int_{\Gamma} \frac{f(\tau)\tau^{\kappa}}{X_1^+(\tau)}d\tau = 0, \quad \kappa = 0, \ldots, |\aleph| - 1. \qquad (2.55)$$

Here, $\aleph := \operatorname{ind} G_1$.

Proof. From Theorem 2.3.3, the assumptions in the theorem are sufficient for the Fredholmness of the associate operator $\mathcal{A}$, and thus equation (2.36) is solvable. From Theorem 2.1.5, the solutions of the boundary problem (2.14) are given by formulas (2.29). The solvability conditions (2.55) are necessary for the solvability of problem (2.14), as stated in Theorem 2.1.5. Finally, from the Sokhotski–Plemelj formulas, we have (2.54). □

2.3.3 $L^\infty(\Gamma)$-factorable coefficients

The Fredholmness of the boundary problem (2.12), for an essentially measurable function admitting a factorisation in $\mathscr{X}(\Gamma)$ as in (2.8), can be characterised through Theorem 2.2.5. In particular, we have the following.

Proposition 2.3.7. *Let Γ be a Lyapunov curve and $\mathscr{X}(\Gamma)$ be any BFS satisfying (2.2)–(2.6). The boundary problem (2.12) is Fredholm iff its coefficient function ab^{-1} admits a factorisation in $\mathscr{X}(\Gamma)$ given in (2.8).*

Note that from Theorem 2.2.6, the invertibility of the coefficient in problem (2.10) is a necessary condition for its normal solvability.

Theorem 2.3.8. *Let* Γ *be a Lyapunov curve and* $\mathscr{X}(\Gamma)$ *be any BFS satisfying assumptions (2.2)–(2.6). Equation (2.36) with essentially bounded coefficients* a *and* b *has solutions if and only if* $a, b \in \mathcal{G}(L^\infty(\Gamma))$ *and* ab^{-1} *admits a factorisation (2.8). The solutions are described as follows, depending on the different situations:*

(a) $(\aleph < 0)$

$$\varphi(t) = (c_+^{-1}(t)t^{-\aleph}P_+ + c_-(t)P_-)c_-^{-1}(t)b^{-1}(t)g(t)$$
$$+ (c_+^{-1}(t)t^{-\aleph} + c_-(t))\rho(t), \tag{2.56}$$

where ρ *is an arbitrary polynomial of degree* $|\aleph| - 1$.
(b) $(\aleph \geq 0)$ *The unique solution in this case is as in (2.56) with* $\rho(t) \equiv 0$. *In addition, it is necessary that*

$$\int_\Gamma c_-^{-1}(\tau)b(\tau)f(\tau)\tau^{-\kappa}d\tau = 0, \quad \kappa = 1, \ldots, \aleph. \tag{2.57}$$

Here, $\aleph := \operatorname{ind} ab^{-1}$.

Proof. From Theorem 2.2.7, equation (2.36) is solvable. Note that conditions (2.30) hold for the factorable function $ab^{-1}\mathcal{G}(L^\infty(\Gamma))$. Therefore, from Theorem 2.1.6, the solutions of the boundary problem (2.12) have the form (2.31). In this last case, from Corollary 2.1.7, for the solvability of the BVP (2.12), (2.57) is necessary; therefore, from the Sokhotski–Plemelj formulas, the general solutions of equation (2.36) have the representation given by (2.56), with an arbitrary polynomial function ρ if $\aleph < 0$ and $\rho(t) \equiv 0$ if $\aleph \geq 0$, in which case, condition (2.57) should be satisfied. See Theorem 2.2.7 and equality (2.40). □

2.4   SIEs with Carleman Shift

Let $\alpha(t)$ be a homeomorphism of Γ onto itself, which may preserve or change the orientation of Γ, and let us suppose that, at every point t, there exist the derivative $\alpha'(t)$ satisfying $\alpha'(t) \neq 0$ and $\alpha'(t)$ satisfying the Hölder condition. In addition, we assume that $\alpha(t)$ satisfies the so-called *Carleman condition*:

$$\alpha^2(t) = (\alpha \circ \alpha)(t) = t. \tag{2.58}$$

Moreover, we assume that for the space $\mathscr{X}(\Gamma)$, we get that

$\alpha(t)$ induces a bounded shift operator $(W\varphi)(t) = \varphi(\alpha(t))$ on $\mathscr{X}(\Gamma)$.

$$(2.59)$$

Note that (2.58) implies that W satisfies the Carleman condition $W^2 = I$ (see, for example, Karapetiants and Samko, 2001; Kravchenko and Litvinchuk, 1994). Among all the different kinds of Carleman shift operators, here we consider those operators W satisfying $WS = \gamma SW$, where $\gamma = \pm 1$. When $\gamma = 1$ (i.e. α preserves the orientation of Γ), W is called a *commutative Carleman shift operator*, and for $\gamma = -1$ (i.e. α reverts the orientation of Γ), it is called an *anti-commutative Carleman shift operator*.

In this section, the solvability of the following class of integral equations is studied in the space $\mathscr{X}(\Gamma)$ over a Lyapunov curve Γ, satisfying (2.2)–(2.4), (2.6), and (2.59):

$$f(t)\varphi(t) + g(t)\frac{1}{\pi i}\text{p.v.}\int_\Gamma \frac{\varphi(\tau)}{\tau - t}d\tau + g(t)\frac{1}{\pi i}\text{p.v.}\int_\Gamma \frac{\varphi(\tau)}{\tau - \alpha(t)}d\tau = h(t),$$

$$(2.60)$$

where the elements $f(t)$ and $g(t)$ are essentially bounded functions, with $f(t) \neq 0$ on Γ and $\alpha(t)$ is a Carleman shift function.

Let us consider the following complementary projection operators on $\mathscr{X}(\Gamma)$:

$$P_1 := \frac{1}{2}(I - W) \quad \text{and} \quad P_2 := \frac{1}{2}(I + W).$$

Note that $W^k = \sum_{j=1}^2 (-1)^{kj} P_j$, $k = 1, 2$, and

$$P_k = \frac{1}{2}\sum_{j=1}^2 (-1)^{k(1-j)} W^{j+1}, \quad k = 1, 2. \tag{2.61}$$

In view of obtaining the solvability results, the following lemma will be useful.

Lemma 2.4.1. *Let Γ be a Lyapunov curve, and let $\mathscr{X}(\Gamma)$ be a BFS satisfying (2.2)–(2.4), (2.6), and (2.59). Let $\psi \in \mathscr{X}(\Gamma)$. Then, for $z \in \mathbb{C}$, we have*

$$(P_k S\psi)(z) = \begin{cases} (SP_k\psi)(z), & \text{if } W \text{ is a commutative shift operator,} \\ (SP_{3-k}\psi)(z), & \text{if } W \text{ is an anti-commutative shift operator.} \end{cases}$$

$$(2.62)$$

Proof. We have directly

$$(P_k S\psi)(z) = \frac{1}{2}\{(S\psi)(z) + (-1)^k W(S\psi)(z)\}$$

$$= \frac{1}{2}\{(S\psi)(z) + (-1)^k \gamma (WS\psi)(z)\},$$

where $\gamma = \pm 1$, depending on whether W is a commutative or anti-commutative Carleman shift operator, respectively. From here, equality (2.62) follows. □

2.4.1 *Auxiliary system of equations and their solvability*

In this section, we discuss the existence and uniqueness of the eventual solutions of equation (2.60). Moreover, we provide explicit representations of such solutions. To this end, in particular, we use the projection methods, as given by Castro and Rojas (2010, 2011), Castro *et al.* (2015), Chuan and Tuan (2003), Chuan *et al.* (2008), and Tuan (1996), so that we will be able to transform the initial equation into a system of equations which can be solved by means of a Riemann BVP technique.

Let us introduce the following functions: for $k = 1, 2$,

$$f_\alpha(t) := f(t)f(\alpha(t)), \tag{2.63}$$

$$[fg]_k(t) := f(\alpha(t))g(t) + (-1)^k f(t)g(\alpha(t)), \tag{2.64}$$

$$[fh]_k(t) := \frac{1}{2}(f(\alpha(t))h(t) + (-1)^k f(t)h(\alpha(t))). \tag{2.65}$$

Note that with the complementary projections P_k ($k = 1, 2$) given in (2.61), the functions in (2.64) and (2.65) can be rewritten as $[fg]_k(t) = 2P_k[f(\alpha(t))g(t)]$ and $[fh]_k(t) = P_k[f(\alpha(t))h(t)]$, respectively.

Now, we replace equation (2.60) with a simpler and equivalent system of equations. First of all, note that with the projection operators P_k ($k = 1, 2$), we can rewrite equation (2.60) as follows:

$$f(t)\varphi(t) + 2g(t)(P_2 S\varphi)(t) = h(t). \tag{2.66}$$

Proposition 2.4.2. *Let $\varphi \in \mathscr{X}(\Gamma)$. Then, φ is a solution of (2.66) if and only if $\{\varphi_k := P_k\varphi, k = 1, 2\}$ is a solution of the following system:*

$$f_\alpha(t)\varphi_k(t) + [fg]_k(t)\,[(S\varphi_2)(t)] = [fh]_k(t)$$

if α preserves orientation, or $\qquad\qquad\qquad$ (2.67)

$$f_\alpha(t)\varphi_k(t) + [fg]_k(t)\,[(S\varphi_1)(t)] = [fh]_k(t), \text{ otherwise,}$$

where $f_\alpha(t)$, $[fg]_k(t)$, and $[fh]_k(t)$ $(k = 1, 2)$ are defined in (2.63), (2.64), and (2.65), respectively.

Proof. Suppose that $\varphi \in \mathscr{X}(\Gamma)$ is a solution of (2.66). Multiplying by $f(\alpha(t))$, we have

$$f(\alpha(t))f(t)\varphi(t) + 2f(\alpha(t))g(t)(P_2 S\varphi)(t) = f(\alpha(t))h(t).$$

Applying the projections P_k $(k = 1, 2)$ to both sides of the above equation, we get

$$P_k[f(\alpha(t))f(t)\varphi](t) + 2P_k[f(\alpha(t))g(t)(P_2 S\varphi)](t) = P_k[f(\alpha(t))h(t)]. \quad (2.68)$$

By using (2.61) and the fact that $WP_2 = P_2$, we can verify that

$$P_k[f(\alpha(t))f(t)\varphi](t) = f(\alpha(t))f(t)(P_k\varphi)(t),$$

$$P_k[f(\alpha(t))g(t)(P_2 S\varphi)](t) = P_k[f(\alpha(t))g(t)](P_2 S\varphi)(t).$$

Therefore, we can rewrite (2.68) as

$$f(\alpha(t))f(t)(P_k\varphi)(t) + 2P_k(f(\alpha(t))g(t))(P_2 S\varphi)(t) = P_k[f(\alpha(t))h(t)]. \quad (2.69)$$

Now, by Lemma 2.4.1, we have that $P_2 S = SP_2$ for W being a commutative Carleman shift operator and that $P_2 S = SP_1$ for W being an anti-commutative one. In this way, we conclude that $(P_1\varphi, P_2\varphi)$ is a solution of (2.67).

Conversely, suppose that there exists φ such that $(P_1\varphi, P_2\varphi)$ is a solution of (2.67). Evidently, Lemma 2.4.1 guarantees that system (2.67) is equivalent to system (2.69); thus, summing k from 1 to 2, we directly obtain, using the last system (2.69), that

$$\sum_{k=1}^{2} [f_\alpha(t)(\varphi_k)(t) + 2P_k[f(\alpha(t))g(t)](P_2 S\varphi)(t)] = \sum_{k=1}^{2} P_k[f(\alpha(t))h(t)]$$

is equivalent to

$$f(\alpha(t))f(t)\varphi(t) + 2f(\alpha(t))g(t)(P_2 S\varphi)(t) = f(\alpha(t))h(t)$$

due to the fact that $f(t) \neq 0$, for $t \in \Gamma$. Then,

$$f(t)\varphi(t) + 2g(t)(P_2 S_\Gamma \varphi)(t) = h(t),$$

which completes the proof. $\qquad\qquad\square$

Proposition 2.4.3. *If (ϕ_1, ϕ_2) is a solution of system* (2.67), *then* $(P_1\phi_1, P_2\phi_2)$ *is also a solution of* (2.67).

Proof. Let (ϕ_1, ϕ_2) be a solution of system (2.67). Applying the projections P_k to both sides of (2.67), we have

$$P_k\left(f_\alpha(t)\phi_k(t) + [fg]_k(t)(S\phi_i)(t)\right) = P_k[fh]_k(t),$$

$k, i = 1, 2$. Note that $P_k[f_\alpha(t)\phi_k](t) = f_\alpha(t)P_k\phi_k(t)$ and

$$P_k([(fg)]_k(t)(S\phi_i))(t)$$
$$= \frac{1}{2}\{[fg]_k(t)(S\phi_i)(t) + (-1)^k[fg]_k(\alpha(t))W(S_\Gamma\phi_i)(t)\}$$
$$= [fg]_k(t)\frac{1}{2}\{(S\phi_i)(t) + W(S\phi_i)(t)\}. \tag{2.70}$$

Equality (2.70) holds because $[fg]_k(t) = (-1)^k[fg]_k(\alpha(t))$.

Since $P_k[fg]_k = [fg]_k$, the right-hand side of equality (2.70) can be rewritten as $P_k([fg]_k(t))P_2(S\phi_i)(t)$. From (2.67), the value of i depends on the commutative or anti-commutative property of the shift operator W; therefore,

$$P_k[[fg]_k(t)(S\phi_i)](t) = P_k([fg]_k(t))(SP_i\phi_i)(t).$$

Finally, note that $P_k([fh]_k)(t) = [fh]_k(t)$. Thus, $(P_1\phi_1, P_2\phi_2)$ is a solution of (2.67).   □

Theorem 2.4.4. *Equation* (2.66) *has solutions in $\mathscr{X}(\Gamma)$ if and only if the equation*

$$f_\alpha(t)\varphi_2(t) + [fg]_2(t)(S\varphi_2)(t) = [fh]_2(t)$$
if W is a commutative Carleman shift operator, or
$$f_\alpha(t)\varphi_1(t) + [fg]_1(t)(S\varphi_1)(t) = [fh]_1(t) \tag{2.71}$$
if W is an anti-commutative one,

has solutions. Moreover, if $\varphi_k(t)$ $(k = 1, 2)$ is a solution of (2.71), *then equation* (2.66) *has a solution, which is given by the formula*

$$\varphi(t) = \begin{cases} \dfrac{h(t) - 2g(t)[(S\varphi_2)(t)]}{f(t)} & \text{if } WS = SW, \text{ or} \\[4mm] \dfrac{h(t) - 2g(t)[(S\varphi_1)(t)]}{f(t)} & \text{if } WS = -SW. \end{cases} \tag{2.72}$$

Proof. Suppose that $\varphi \in \mathscr{X}(\Gamma)$ is a solution of equation (2.66). By Proposition 2.4.2, we know that $(P_1\varphi, P_2\varphi)$ is a solution of system (2.67). Hence, for W preserving the orientation of Γ, $P_2\varphi$ is a solution of (2.71) and $P_1\varphi$ is the corresponding solution for W reverting the orientation of Γ.

Conversely, suppose that φ_2 is a solution of (2.71). Without loss of generality, we assume now that we are in the preserving orientation case (since the reverting orientation one is dealt with similarly). In this case, system (2.67) has a solution (φ_1, φ_2) determined by

$$\varphi_1(t) = \frac{[fh]_1(t) - [fg]_1(t)[(S\varphi_2)(t)]}{f_\alpha(t)}. \tag{2.73}$$

From Proposition 2.4.3, we have that $P_i\varphi_i$ is also a solution of (2.71), so we have that $(P_1\varphi_1, P_2\varphi_2)$ is also a solution of (2.67). Set

$$\varphi = \sum_{k=1}^{2} P_k\varphi_k. \tag{2.74}$$

It is clear that $P_k\varphi = P_k\varphi_k$. This means $(P_1\varphi, P_2\varphi)$ is a solution of (2.68). From Proposition 2.4.2, it follows that φ is a solution of (2.67). Moreover, from (2.73) and (2.74), we obtain

$$\varphi(t) = \sum_{k=1}^{2} P_k \left[\frac{[fh]_k(t) - [fg]_k[(S\varphi_2)(t)]}{f_\alpha(t)} \right]. \tag{2.75}$$

As before, we can see that

$$\sum_{k=1}^{2} P_k[fh]_k(t) = f(\alpha(t))h(t),$$

$$\sum_{k=1}^{2} P_k([fg]_k(t)[(S\varphi_2)(t)]) = 2f(\alpha(t))g(t)[(S\varphi_2)(t)].$$

Thus, substituting these in (2.75), we have

$$\varphi(t) = \frac{h(t) - 2g(t)[(S\varphi_2)(t)]}{f(t)}. \qquad \square$$

2.4.2   *Closed-form representation of the solutions*

At this point, we have that formula (2.72) gives a representation for the solutions of equation (2.60). In order to obtain a closed-form solution of equation (2.60), we must compute $(S\varphi_k)(t)$, $k = 1, 2$, on formula (2.72).

We can use the Riemann problems associated with equation (2.71) to describe the form of that solution in the cases where the coefficients of the equation are continuous, piecewise continuous, or essentially bounded factorable functions.

For continuous coefficients, let us assume that $f_\alpha \pm [fg]_k \in \mathcal{G}(C(\Gamma))$ $(k = 1, 2)$ and set

$$G(t) = \begin{cases} \dfrac{f_\alpha(t) - [fg]_2(t)}{f_\alpha(t) + [fg]_2(t)} & \text{if } W \text{ commutes,} \\[4mm] \dfrac{f_\alpha(t) - [fg]_1(t)}{f_\alpha(t) + [fg]_1(t)} & \text{if } W \text{ anti-commutes,} \end{cases} \tag{2.76}$$

and

$$H(t) = \begin{cases} \dfrac{[fh]_2(t)}{f_\alpha(t) + [fg]_2(t)} & \text{if } W \text{ commutes,} \\[4mm] \dfrac{[fh]_1(t)}{f_\alpha(t) + [fg]_1(t)} & \text{if } W \text{ anti-commutes.} \end{cases} \tag{2.77}$$

With the functions F and H, equation (2.71) can be rewritten as

$$P_+\varphi_k(t) + G(t)P_-\varphi_k(t) = H(t),$$

which is then reduced to the problem

$$\Psi_k^+(t) = G(t)\Psi_k^-(t) + H(t), \quad G \in \mathcal{G}(C(\Gamma)).$$

Moreover, from the Sokhotski–Plemelj formulas, we have that $(S\varphi_k)(t) = \Psi_k^+(t) + \Psi_k^-(t)$; thus, the representations of the solutions of equations (2.71) for a commuting or anti-commuting shift with continuous coefficients are given in the following result.

Theorem 2.4.5. *Let Γ be a Lyapunov curve, and let $\mathscr{X}(\Gamma)$ be a BFS satisfying (2.2)–(2.4), (2.6), and (2.59). Let $G(t)$ and $H(t)$ be as in (2.76) and (2.77). respectively. Then. equation (2.60) has solutions in $\mathscr{X}(\Gamma)$, and they are given by*

$$\varphi(t) = \frac{h(t) - 2g(t)(S\varphi_k)(t)}{f(t)}, \quad k = 1, 2,$$

where $k = 1$ in the case when W is a commutative Carleman shift operator and $k = 2$ if W is an anti-commutative Carleman shift operator. In addition, for computing $(S\varphi_k)(t) = \Psi_k^+(t) + \Psi_k^-(t)$, we have the following different situations:

(a) ($\aleph \geq 0$) *In this case, we have*

$$\Psi_k^+(t) = \frac{e^{h(t)}}{2\pi i} \int_\Gamma \frac{H(\tau)}{e^{h(\tau)}} \frac{d\tau}{\tau - t} + e^{h(t)}\rho(t), \tag{2.78}$$

$$\Psi_k^-(t) = \frac{(t-z_0)^\aleph e^{h(t)}}{2\pi i} \int_\Gamma \frac{H(\tau)}{e^{h(\tau)}} \frac{d\tau}{\tau - t} + (t-z_0)^\aleph e^{h(t)}\rho(t), \tag{2.79}$$

where

$$h(t) = \frac{1}{2\pi} \int_\Gamma \frac{\ln G(\tau)(\tau - z_0)^\aleph}{\tau - t} d\tau, \quad z_0 \in D^+,$$

and $\rho(t) = a_{\aleph-1}t^{\aleph-1} + a_{\aleph-2}t^{\aleph-2} + \cdots + a_0$.

(b) ($\aleph < 0$) *For this case, the solution is unique, and* $\Psi_k^\pm$ *are as in* (2.78) *and* (2.79) *with* $\rho(t) \equiv 0$. *In addition, it is necessary that*

$$\int_\Gamma \frac{H(\tau)\tau^\kappa}{e^{h(\tau)}} = 0, \quad \kappa = 0, \ldots, |\aleph| - 1.$$

Proof. From Theorem 2.4.4, we know that equation (2.60) has solutions if and only if equation (2.71) has solutions. Furthermore, the solutions of (2.60) are given by (2.72). Thus, we compute the solutions of equation (2.71). For this, we use the corresponding Riemann BVP associated with equation (2.71). Namely, by means of the Sokhotski–Plemelj formulas, equation (2.71) is reduced to the following boundary problem: find a sectionally analytic function $\Psi_k(z)$ ($\Psi_k(z) = \Psi_k^+(z)$ for $z \in D^+$, $\Psi_k(z) = \Psi_k^-(z)$ for $z \in D^-$) vanishing at infinity and satisfying the condition

$$\Psi_k^+(t) = G(t)\Psi_k^-(t) + H(t) \tag{2.80}$$

on the boundary Γ, where the functions $G(t)$ and $H(t)$ are defined in (2.76) and (2.77), respectively.

From Theorem 2.1.2, the solutions of problem (2.80) read as follows:

(1) **Case $\aleph \geq 0$:** In this case, the solutions are given by (cf. formula (2.16))

$$\Psi_k^\pm(t) = \frac{X^\pm(t)}{2\pi i} \int_\Gamma \frac{H(\tau)}{X^+(\tau)} \frac{d\tau}{\tau - z} + X^\pm(t)\rho(t), \tag{2.81}$$

where $X^+(t) = \exp h(t)$, $X^-(t) = (t-z_0)^\aleph \exp h(t)$ ($z_0 \in D^+$), and

$$h(t) = K\left(\ln G(\tau)(\tau - z_0)^\aleph\right)(t).$$

ρ is an arbitrary polynomial of degree $\aleph - 1$. The second item on the right-hand side of formulas (2.81) is the general solution of the homogeneous $(H(t) \equiv 0)$ Riemann problem (2.80), while the first item is a particular solution of the corresponding non-homogeneous problem (2.80).

(2) **Case $\aleph < 0$:** For this case, $\Psi_k^{\pm}$ are as in (2.81), and $\rho(z) \equiv 0$. In addition, it is necessary that

$$\int_\Gamma \frac{H(\tau)\tau^\kappa}{e^{h(\tau)}} d\tau = 0, \quad \kappa = 0, \ldots, |\aleph| - 1.$$

This completes the proof. □

For the case when equation (2.60) has essentially bounded coefficients, let us assume $f_\alpha \pm [fg]_k \in \mathcal{G}(L^\infty(\Gamma))$ $(k = 1,2)$ and define the functions

$$\widetilde{G}(t) = \frac{1}{G(t)} \tag{2.82}$$

for the function G given in (2.76) and

$$\widetilde{H}(t) = \begin{cases} \dfrac{[fh]_2(t)}{f_\alpha(t) - [fg]_2(t)} & \text{if } W \text{ commutes,} \\[3mm] \dfrac{[fh]_1(t)}{f_\alpha(t) - [fg]_1(t)} & \text{if } W \text{ anti-commutes.} \end{cases} \tag{2.83}$$

The functions $\widetilde{F}$ and $\widetilde{H}$ allow us to rewrite equation (2.71) as

$$\widetilde{G}(t)P_+\varphi_k(t) + P_-\varphi_k(t) = \widetilde{H}(t). \tag{2.84}$$

Theorem 2.4.6. *Let Γ be a Lyapunov curve, and let $\mathscr{X}(\Gamma)$ be a BFS satisfying (2.2)–(2.6) and (2.59). Let $\widetilde{G}(t)$ and $\widetilde{H}(t)$ be as in (2.82) and (2.83), respectively; moreover, let us suppose that $\widetilde{G}$ admits a factorisation in $\mathscr{X}(\Gamma)$ and $G_-(t)t^\aleph G_+(t)$. Then, equation (2.60) has solutions in $\mathscr{X}(\Gamma)$, which are given by*

$$\varphi(t) = \frac{h(t) - 2g(t)(S\varphi_k)(t)}{f(t)}, \quad k = 1,2,$$

where $k = 1$ in the case when W is a commutative Carleman shift operator and $k = 2$ if W is an anti-commutative Carleman shift operator. In addition, for computing $(S\varphi_k)(t) = \Psi_k^+(t) + \Psi_k^-(t)$, we have the following different situations:

52   *Approximation and Regularisation Methods for Operator-Functional Equations*

(a) ($\aleph < 0$) *In this case, we have*

$$\Psi_k^+(t) = G_+^{-1}(t)t^{-\aleph}P_+G_-^{-1}(t)H(t) + G_+^{-1}(t)t^{-\aleph}\rho(t), \qquad (2.85)$$

$$\Psi_k^-(t) = G_-(t)P_-G_-^{-1}(t)H(t) + G_-(t)\rho(t), \qquad (2.86)$$

where $\rho(t) = a_{-\aleph-1}t^{-\aleph-1} + a_{-\aleph-2}t^{-\aleph-2} + \cdots + a_0$.

(b) ($\aleph \geq 0$) *For this case,* $\Psi_k^\pm$ *are as in* (2.85) *and* (2.86) *with* $\rho(t) \equiv 0$. *In addition, it is necessary that*

$$\int_\Gamma G_-^{-1}(\tau)(f_\alpha(\tau) - [fg]_k(\tau))[fh]_k(\tau)\tau^{-\kappa} = 0, \quad \kappa = 1, \dots, \aleph, \ k = 1, 2.$$

Proof. From Theorem 2.4.4, we know that equation (2.60) has solutions if and only if equation (2.71) has solutions. Furthermore, the solutions of (2.60) are given by (2.72). Thus, we compute the solutions of equation (2.71). For this, we use the corresponding Riemann BVP associated with equation (2.84): find a sectionally analytic function vanishing at infinity and satisfying the condition

$$\Psi_k^-(t) - G(t)\Psi_k^+(t) = -H(t) \qquad (2.87)$$

on the boundary Γ, where the functions $\widetilde{G}(t)$ and $\widetilde{H}(t)$ are defined in (2.82) and (2.83), respectively. Since $\widetilde{G}(t)$ admits a factorisation in $\mathscr{X}(\Gamma)$, as in Section 2.3, we are able to use Theorem 2.1.6. Thus, the solutions of the boundary problem (2.87) read as follows:

(1) **Case $\aleph < 0$:** In this case, the solutions are given by

$$\Psi_k^+(t) = G_+^{-1}t^{-\aleph}P_+G^{-1}(t)H(t) + G_+^{-1}(t)t^{-\aleph}\rho(t), \qquad (2.88)$$

$$\Psi_k^-(t) = G_-(t)P_-G_-^{-1}(t)H(t) + G_-(t)\rho(t), \qquad (2.89)$$

where $\rho(t) = a_{-\aleph-1}t^{-\aleph-1} + a_{-\aleph-2}t^{-\aleph-2} + \cdots + a_0$. The second item on the right-hand side of formulas (2.88) and (2.89) is the general solution of the homogeneous ($H(t) \equiv 0$) Riemann problem (2.87), while the first item is a particular solution of the corresponding non-homogeneous problem (2.87).

(2) **Case $\aleph \geq 0$:** For this case, $\Psi_k^\pm$ are as in (2.88) and (2.89), and $\rho(z) \equiv 0$. In addition, it is necessary that

$$\int_\Gamma G_-^{-1}(\tau)(f_\alpha(\tau) - [fg]_k(\tau))[fh]_k(\tau)\tau^{-\kappa} = 0, \quad \kappa = 1, \dots, \aleph, \ k = 1, 2.$$

If $\aleph = 0$, then problem (2.87) has a unique solution. This completes the proof. $\qquad\square$

2.4.3   The Fredholmness of the singular integral operator with shift

Note that in the operator theory approach, equation (2.60) is associated with the singular integral operator

$$S := fI + gS + gWS : \mathscr{X}(\Gamma) \longrightarrow \mathscr{X}(\Gamma).$$

The projection method, which we used earlier, allows us to establish a Fredholm criterion for the operator S on $\mathscr{X}(\Gamma)$ by means of a non-explicit equivalence operator relation.

Theorem 2.4.7. *Let* Γ *be a Lyapunov curve, and let* $\mathscr{X}(\Gamma)$ *be a BFS satisfying* (2.2)–(2.4), (2.6) *and* (2.59). *Then, the operator* $S := fI + gS + gWS$ *is a* Φ *operator on* $\mathscr{X}(\Gamma)$ *if and only if* $(f_\alpha - [fg]_k)(f_\alpha + [fg]_k)^{-1} \in \mathcal{G}(C(\Gamma))$. *The functions* f_α *and* $[fg]_k$ *are given in* (2.63) *and* (2.64), *respectively,* $k = 1$ *if* W *anti-commutes and* $k = 2$ *if* W *commutes. Moreover, under the presence of the Fredholm property,* $\mathrm{Ind}\,S = \mathrm{ind}(f_\alpha - [fg]_k)$ $(f_\alpha + [fg]_k)^{-1} =: \aleph.$

Proof. From Theorem 2.4.4, we know that equation (2.60) is solvable if and only if the equation

$$f_\alpha(t)\varphi_k(t) + [fg]_k(t)(S\varphi_k)(t) = [fh]_k(t) \quad (k = 1, 2) \tag{2.90}$$

is solvable, where f_α, $[fg]_k(t)$, and $[fh]_k(t)$ are given in (2.63), (2.64), and (2.65), respectively, and $k = 1$ or $k = 2$ depending on the commutative nature of the shift operator W. Moreover, from formula (2.72), the dimension of the set of solutions of equations (2.60) and (2.90) coincide.

On the other hand, by using the functions G and H defined in (2.82) and (2.83), respectively, equation (2.90) can be rewritten as

$$P_+\varphi_k(t) + G(t)P_-\varphi_k(t) = H(t).$$

Therefore, the regularity properties of the operators $P_+ + GP_-$ and S coincide.

Finally, from Theorem 2.2.4, the operator $P_+ + GP_-$ is a Φ operator with Fredholm index $\aleph = \mathrm{ind}\,G$ if and only if $G \in \mathcal{G}(C(\Gamma))$, and so this is transferred to the operator S, i.e. we conclude that $f_\alpha \pm [fg]_k \in \mathcal{G}(C(\Gamma))$ (with $\mathrm{ind}\,G = \aleph$) if and only if S is a Φ operator with $\mathrm{Ind}\,S = \aleph$. $\qquad\square$

In a similar way, the Fredholmness of the operator S with piecewise continuous and factorable essentially bounded functions as coefficients can be proved by using Theorems 2.3.3 and 2.2.5, respectively.

Theorem 2.4.8. *Let Γ be a Lyapunov curve, and let $\mathscr{X}(\Gamma)$ be a BFS satisfying (2.2)–(2.4), (2.6), (2.59), and Axioms 1 and 2. Then, the operator $S := fI + gS + gWS$ is a Φ operator on $\mathscr{X}(\Gamma)$ if $f_\alpha \pm [fg]_k \in \mathcal{G}(PC(\Gamma))$, and the function $(f_\alpha + [fg]_k)(f_\alpha - [fg]_k)^{-1}$ is $\mathscr{X}(\Gamma)$-nonsingular with discontinuity points $t_1, \ldots, t_n$. Moreover, we suppose that the curve Γ at the points t_k has at least one-sided tangents. The functions f_α and $[fg]_k$ are given in (2.63) and (2.64), respectively, $k = 1$ if W anti-commutes, and $k = 2$ if W commutes. In this case,*

$$\operatorname{Ind} S = \operatorname{ind}(f_\alpha + [fg]_k)(f_\alpha - [fg]_k)^{-1}.$$

The condition $f_\alpha \pm [fg]_k \in \mathcal{G}(PC(\Gamma))$ is also necessary for the Fredholmness of the operator S. On the other hand, if the index functions α and β of the space $\mathscr{X}(\Gamma)$ coincide at the points t_k of discontinuity of the coefficients $(f_\alpha + [fg]_k)(f_\alpha - [fg]_k)^{-1}$, then the $\mathscr{X}(\Gamma)$-nonsingularity of $(f_\alpha + [fg]_k)(f_\alpha - [fg]_k)^{-1}$ is necessary as well.

Theorem 2.4.9. *Let Γ be a Lyapunov curve, and let $\mathscr{X}(\Gamma)$ be a BFS satisfying (2.2)–(2.6) and (2.59). Then, the operator $S := fI + gS + gWS$ is a Φ operator on $\mathscr{X}(\Gamma)$ if and only if the functions $f_\alpha \pm [fg]_k \in \mathcal{G}(L^\infty(\Gamma))$ and $\widetilde{G} = (f_\alpha + [fg]_k)(f_\alpha - [fg]_k)^{-1}$ admit a factorisation (2.8) in $\mathscr{X}(\Gamma)$ with $\operatorname{ind} \widetilde{G} = \aleph$. The functions f_α and $[fg]_k$ are given in (2.63) and (2.64), respectively, $k = 1$ if W anti-commutes, and $k = 2$ if W commutes. Moreover, under the presence of the Fredholm property, $\operatorname{Ind} S = -\aleph$.*

2.4.4   *The case of variable Lebesgue spaces*

Variable Lebesgue spaces are one of the most well-known Banach function spaces (BFSs). Their fundamental study has been growing rapidly since the past three decades, apart from mathematical curiosity, for possible applications to image restoration and to models with the so-called non-standard local growth in fluid mechanics, elasticity theory, and differential equations; see, for instance, the work of Chen *et al.* (2006), Diening and Ružička (2003), and the references therein.

In this section, we show that all the results given in this chapter are valid in the framework of variable Lebesgue spaces. To show this, we display that conditions (2.2)–(2.6), (2.59), and Axioms 1 and 2 imposed on $\mathscr{X}(\Gamma)$ are, in fact, well-established results on variable Lebesgue spaces.

The space $L^{p(\cdot)}(\Gamma)$ over a Jordan curve Γ of finite length ℓ is defined as the set of all measurable complex-valued functions f on Γ such that

$I_p(\lambda f) < \infty$, for some $\lambda = \lambda(f) > 0$, where

$$I_p(f) = \int_\Gamma |f(t)|^{p(t)} |dt| = \int_0^\ell |f(t(s))|^{p(t(s))} ds.$$

This set becomes a Banach space with respect to the (Luxemburg) norm

$$\|f\|_{p(\cdot)} := \inf \left\{ \lambda > 0 \ : \ I_p\left(\frac{f}{\lambda}\right) \le 1 \right\}.$$

For a study of the fundamental properties of these spaces, we refer the reader to Cruz-Uribe and Fiorenza (2013), Diening *et al.* (2011), and Rafeiro and Rojas (2014).

Let us assume that $p : \Gamma \longrightarrow [1, \infty)$ is a measurable function with the condition

$$1 < p_- = \text{ess inf}\, p(t) \le p(t) \le p_+ = \text{ess sup}\, p(t) < \infty, \quad t \in \Gamma. \tag{2.91}$$

In the sequel, we need the following condition on $p(t)$:

$$|p(t_1) - p(t_2)| \le \frac{A}{-\ln|t_1 - t_2|}, \ |t_1 - t_2| \le \frac{1}{2}, \ t_1, t_2 \in \Gamma, \tag{2.92}$$

where $A > 0$ does not depend either on t_1 and t_2 or on the function $p_*(s) = p(t(s))$:

$$|p_*(s_1) - p_*(s_2)| \le \frac{A}{-\ln|s_1 - s_2|}, \ |s_1 - s_2| \le \frac{1}{2}, \ s_1, s_2 \in [0, \ell]. \tag{2.93}$$

Since $|t(s_1) - t(s_2)| \le |s_1 - s_2|$, condition (2.92) always implies (2.93). Inversely, (2.93) implies (2.92) if there exists $\lambda > 0$ such that $|s_1 - s_2| \le c|t(s_1) - t(s_2)|^\lambda$, with some $c > 0$. Therefore, conditions (2.92) and (2.93) are equivalent on Jordan curves. Moreover, it is valid on general curves satisfying the so-called chord condition.

2.4.5   *On the suitableness of $L^{p(\cdot)}(\Gamma)$ for the results*

In order to establish the validity of assumptions (2.2)–(2.6) and (2.59), as well as Axioms 1 and 2, on the spaces $L^{p(\cdot)}(\Gamma)$, we assume that the properties of the exponent $p(t)$ (2.91)–(2.93) hold:

(2.2) $C(\Gamma) \subset L^{p(\cdot)}(\Gamma) \subset L^1(\Gamma)$. This follows from (2.91).

(2.3) $\|af\|_{L^{p(\cdot)}(\Gamma)} \le \sup_{t \in \Gamma} |a(t)| \cdot \|f\|_{L^{p(\cdot)}(\Gamma)}$, for $a \in L^\infty(\Gamma)$, which is evident.

(2.4) The operator S is bounded in $L^{p(\cdot)}(\Gamma)$. This fact is proved by Kokilashvili and Samko (2003b, Theorem 2).

(2.5) $L^{p(\cdot)}(\Gamma)$ is reflexive. This is proved by Kovacik and Rakosnik (1991, Corollary 2.7).

(2.6) $C^\infty(\Gamma)$ is dense in $L^{p(\cdot)}(\Gamma)$. This is given by Kokilashvili and Samko (2003b, Theorem 4.1).

(2.59) $\alpha(t)$ induces a bounded shift operator $(W\varphi)(t) = \varphi(\alpha(t))$ on $L^{p(\cdot)}(\Gamma)$. In fact, for a shift function α as in p. 43 and an exponent function p satisfying (2.91)–(2.93), Paatashvili (2010, Lemma 2) proved that the functions $p_\alpha(t) := p(\alpha(t))$ and $\overline{p}_\alpha(t) := \max(p(t), p_\alpha(t))$ satisfy (2.91)–(2.93) as well. Also, Paatashvili (2010) showed that $L^{p(\cdot)}(\Gamma) \cap L^{p_\alpha(\cdot)}(\Gamma) = L^{\overline{p}_\alpha(\cdot)}(\Gamma)$; therefore, the operators W and S are bounded on the space $L^{\overline{p}_\alpha(\cdot)}(\Gamma)$.

Axiom 1 in the space $L^{p(\cdot)}(\Gamma)$ was proved by Kokilashvili and Samko (2003b, Theorem 2). For Axiom 2, the embedding $L^{p(\cdot)}(\Gamma, |t-t_0|^\gamma) \subset L^1(\Gamma)$ if $\gamma < \frac{1}{q(t_0)}$ follows from the Hölder inequality on $L^{p(\cdot)}(\Gamma)$, and the denseness of $C^\infty(\Gamma)$ in the space $L^{p(\cdot)}(\Gamma, |t - t_0|^\gamma)$, for $t_0 \in \Gamma$, is a particular case of Theorem 4.1 given by Kokilashvili and Samko (2003b).

On the other hand, the dual space of $L^{p(\cdot)}(\Gamma)$ is $L^{p'(\cdot)}(\Gamma)$ (Corollary 2.7 in Kovacik and Rakosnik, 1991), where $p'(t) = \frac{p(t)}{p(t)-1}$. Corollary 2.12 by Kovacik and Rakosnik (1991) asserts that $L^{p(\cdot)}(\Gamma)$ is separable. Then, the adjoint operator of S is well defined in $L^{p'(\cdot)}(\Gamma)$. The denseness of the rational functions on the variable Lebesgue spaces is given in Theorem 4.1 by Kokilashvili and Samko (2003b). Then, the complementary projections $P_\pm$ are well defined, as are the subspaces

$$L_+^{p(\cdot)}(\Gamma) := P_+ L^{p(\cdot)}(\Gamma), \quad \overset{\circ}{L_-^{p(\cdot)}}(\Gamma) := P_- L^{p(\cdot)}(\Gamma),$$

$$L_-^{p(\cdot)}(\Gamma) := \overset{\circ}{L_-^{p(\cdot)}}(\Gamma) \dotplus \mathbb{C}.$$

Now, we can introduce a factorisation for an invertible function $a \in L^\infty(\Gamma)$ in $L^{p(\cdot)}(\Gamma)$. A function $a \in \mathcal{G}(L^\infty(\Gamma))$ admits a *factorisation in* $L^{p(\cdot)}(\Gamma)$ if it can be written in the form

$$a(t) = a_-(t)t^\aleph a_+(t), \quad \text{a.e. on } \Gamma,$$

where $\aleph \in \mathbb{Z}$ and:

(i) $a_- \in L_-^{p(\cdot)}(\Gamma)$, $a_-^{-1} \in L_-^{p'(\cdot)}(\Gamma)$, $a_+ \in L_+^{p'(\cdot)}(\Gamma)$, and $a_+^{-1} \in L_+^{p(\cdot)}(\Gamma)$;
(ii) the operator $a_+^{-1} S a_+ I$ is bounded in $L^{p(\cdot)}(\Gamma)$.

The integer $\aleph$ is referred to as the *index of the function* a and is denoted by ind a. We can prove that the number $\aleph$ is uniquely determined.

We would like to point out that, for this case, we can use the Smirnov class of functions $E^p(D)$ instead of the class $E^1(D)$ because Theorem 3.3 by Kokilashvili *et al.* (2005) assures us that if S is bounded from $L^{p(\cdot)}(\Gamma)$ to $L^p(\Gamma)$ $(1 < p < \infty)$, then for arbitrary $\varphi \in L^{p(\cdot)}(\Gamma)$, the corresponding analytic function of a Cauchy-type integral whose non-tangential limits are φ belongs to $E^p(D)$.

Using this factorisation and due to the fact that $L^{p(\cdot)}(\Gamma)$ satisfies all the assumptions imposed on the space $\mathscr{X}(\Gamma)$, all the results given in the previous sections can be validated, with obvious modifications, for this case.

Chapter 3

Approximation Methods for Linear Operator Equations and Nonlinear Integral Equations

3.1 On Perturbation Method for Equations of the First Kind: Regularisation and Application

Let A be a bounded linear operator in a Banach space X with range $R(A)$ in another Banach space Y. Let us consider the following operator equation:

$$Ax = f, \quad f \in R(A). \tag{3.1}$$

We assume that $R(A)$ can be non-closed, i.e. an equation of the first kind, and $\operatorname{Ker} A \neq \{0\}$. In many practical problems, one is required to solve an approximate equation,

$$\tilde{A}x = \tilde{f}, \tag{3.2}$$

instead of an exact equation. Here, $\tilde{A}$ and $\tilde{f}$ are approximations of the exact operator A and the right-hand side function f, respectively, such that

$$||\tilde{A} - A|| \leq \delta_1, \quad ||\tilde{f} - f|| \leq \delta_2, \quad \delta = \max\{\delta_1, \delta_2\}. \tag{3.3}$$

The problem of solving equation (3.2) is ill-posed. Therefore, the equation is unstable and needs regularisation in most real-world applications. The fundamental results in regularisation theory and the various methods for solving inverse problems emerged from the scientific schools of A. N. Tikhonov, V. I. Ivanov, and M. M. Lavrentiev.

Many regularisation methods have been proposed for operator equation (3.1), with some of the most efficient being Tikhonov's method of stabilising functional, the quasi-solution method suggested by Ivanov, the Lavrentiev perturbation method, and V. A. Morozov's discrepancy principle. Variational approaches, spectral theory, perturbation theory, and functional analysis methods play a principal role in regularisation theory. For a comprehensive introduction to the theory of ill-posed problems and linear operators, readers may refer to the monograph by Tikhonov and Arsenin (1995) and the textbook by Lavrentiev and Savel'ev (1995). V. P. Maslov established the equivalence of solving ill-posed problems to the existence and convergence of the regularisation process. It is to be noted that not all ill-posed problems have solutions and therefore can be regularised. When it comes to the Hadamard and Tikhonov well-posedness conditions, one can see that Tikhonov well-posedness can be achieved by narrowing the initial space to the well-posedness space. Therefore, a problem that is well-posed according to Tikhonov (which may be ill-posed according to Hadamard) is often called a conditionally well-posed problem (classified as an equation of the second kind).

It is worth noting that there has been constant interest in regularisation methods for their application in interdisciplinary research related to signal and image processing, machine learning, artificial intelligence, and various inverse problems.

3.1.1   *Introduction*

In this section, we address the construction of regularised processes by introducing the following perturbed (regularised) equation:

$$Ax_\alpha + B(\alpha)x_\alpha = f. \tag{3.4}$$

In this section, we follow the work of Sidorov and Trenogin (1981), and by regularisation, we understand the following: a solution of the regularised equation (3.4) converges to one of the solutions of the given equation (3.1) when error δ tends to zero, if such a solution exists. We call the following solution of the regularised equation (3.4) a *B-normal solution.*

It is to be noted that a regularisation method based on the perturbed equation was first proposed by Lavrentiev for completely continuous self-adjoint and positive operators, A and $B(\alpha) \equiv \alpha$.

Remark 3.1.1. If a given equation is itself not solvable but a pseudo-solution satisfies it, then the regularised equation is supposed to have a solution that converges to a pseudo-solution when the error level δ tends to zero. A pseudo-solution is defined as an element with a minimum norm that provides a discrepancy minimum norm.

Following Sidorov *et al.* (2002) and Sidorov and Trenogin (1981), we select the operator $B(\alpha)$ to arrive at a solution x_α that is unique and provides computational stability. We call such an operator *stabilising operator* (SO), and the vector parameter α is called *regularisation parameter*, $\alpha \in S \subset \mathbb{R}^n$. Here, S is an open set, with zero belonging to the boundary of this set (briefly, the S-sectoral neighbourhood of zero in $\mathbb{R}^n$), $\lim_{S \ni \alpha \to 0} B(\alpha) = 0$. The parameter α should be adjusted according to the error level δ of measurements.

Previously, we addressed only the simple case of $B(\alpha) = B_0 + \alpha B_1$, $\alpha \in \mathbb{R}^+$ with Fredholm operator A in equation (3.4). It is to be noted that such an SO with parameter $\alpha \in \mathbb{R}^1$ has been introduced and employed in previous studies (Sidorov and Trenogin, 1976, 1981). Here operator B_0 was selected according to the Schmidt lemma (Vainberg and Trenogin, 1964).

Such an SO has been employed in the development and justification of iterative methods of calculating Fredholm points λ_0, finding zeros, and determining the elements of the generalised Jordan sets of operator functions. It has also been used for the construction of approximate methods in the theory of branching of solutions of nonlinear operator equations with parameters and the construction of solutions for differential-operator equations with irreversible operator coefficients in the main part. In this monograph, we present a novel theory for operator system regularisation.

The section is organised as follows. First, the sufficient conditions are derived for the case when perturbed equation (3.4) enables a regularisation process. Next, we discuss how the classic Banach–Steinhaus theorem plays an important role in the choice of SO $B(\alpha)$. Finally, the application of the regularising equation of the form (3.4) to the problem of stable differentiation is described.

3.1.2 *The fundamental theorem of regularisation using the perturbation method*

Apart from equations (3.1), (3.2), and (3.4), let us introduce the following equations:

$$(Ax + B(\alpha))x = \tilde{f}, \tag{3.5}$$

$$(\tilde{A}x + B(\alpha))x = \tilde{f}. \tag{3.6}$$

The errors in the operator $B(\alpha)$ can always be included in the operator $\tilde{A}$. Equation (3.6) is called *regularised equation* (RE) for problem (3.2). The following estimates are assumed to be fulfilled:

$$\|(A + B(\alpha))^{-1}\| \leq c(|\alpha|), \tag{3.7}$$

$$\|B(\alpha)\| \leq d(|\alpha|), \tag{3.8}$$

where $c(|\alpha|)$ is a continuous function, $\alpha \in S \subset \mathbb{R}^n$, $0 \in \overline{S}$, $\lim_{|\alpha| \to 0} c(|\alpha|) = \infty$, $\lim_{|\alpha| \to 0} d(|\alpha|) = 0$. If x^* is a solution to equation (3.1), then $(A + B(\alpha))^{-1} f - x^* = -(A + B(\alpha))^{-1} B(\alpha) x^*$. Therefore, we have the following.

Lemma 3.1.2. *Let x^* be some solution to equation (3.1) and $x(\alpha)$ satisfy equation (3.4). Then, in order that $x_\alpha \to x^*$ for $S \ni \alpha \to 0$, it is necessary and sufficient that the following equality is fulfilled:*

$$S(\alpha, x^*) = ||(A + B(\alpha))^{-1} B(\alpha) x^*|| \to 0 \text{ for } S \ni \alpha \to 0. \qquad (3.9)$$

Definition 3.1.1. We refer to condition (3.9) as *stabilisation condition*, the operator $B(\alpha)$ as *stabilisation operator* if it satisfies condition (3.9), and solution x^* as the *B-normal solution* of equation (3.1).

Remark 3.1.3. Obviously, the limit of the sequence $\{x_\alpha\}$ is unique in normed space; therefore, equation (3.1) can have only one B-normal solution.

From estimates (3.7) and (3.8), we arrive at the following.

Lemma 3.1.4. *Let x_α and $\hat{x}_\alpha$ be solutions of equations (3.4) and (3.5), respectively. If the parameter $\alpha = \alpha(\delta) \in S$ is selected such that $\delta \to 0$, as well as*

$$|\alpha(\delta)| \to 0 \quad \text{and} \quad \delta c(|\alpha(\delta)|) \to 0, \qquad (3.10)$$

then $\lim_{\delta \to 0} ||x_\alpha - \hat{x}_\alpha|| = 0$.

Definition 3.1.2. We call condition (3.10) the coordination condition of the vector parameter α with error level δ.

The coordination condition plays a principal role in all regularisation methods for ill-posed problems. Here, it is assumed to be fulfilled. We also assume that α depends on δ, but for the sake of brevity we omit this fact.

Lemma 3.1.5. *Let estimates (3.7) and (3.8) be satisfied, as is the coordination condition for the regularisation parameter (3.10). Next, we select $q \in (0, 1)$ and find $\delta > 0$ such that, for $\delta \le \delta_0$, the following inequality is fulfilled:*

$$\delta c(|\alpha|)) \le q. \qquad (3.11)$$

Then, $\tilde{A} + B(\alpha)$ is a continuously invertible operator, and the following estimates are fulfilled:

$$\|(\tilde{A} + B(\alpha))^{-1}\| \leq \frac{\|(A + B(\alpha))^{-1}\|}{1 - q}, \tag{3.12}$$

$$\|(\tilde{A} + B(\alpha))^{-1}f\| \leq \|(A + B(\alpha))^{-1}f\| + \delta \frac{c(|\alpha|)}{1 - q}\|(A + B(\alpha))^{-1}f\|. \tag{3.13}$$

Proof. Based on estimate (3.3) for all f, we have

$$\|(\tilde{A} - A)(A + B(\alpha))^{-1}f\| \leq \delta\|(A + B(\alpha))^{-1}f\|. \tag{3.14}$$

Hence, taking into account the estimates (3.7), (3.8), and (3.11), we have the following inequality:

$$\|(\tilde{A} - A)(A + B(\alpha))^{-1}f\| \leq \delta c(|\alpha|) \leq q\|f\|. \tag{3.15}$$

Now, since $q \leq 1$, we have $\tilde{A} + B(\alpha) = (I + (\tilde{A} - A)(A + B(\alpha))^{-1})(A + B(\alpha))$. Then, the existence of the inverse operator $(\tilde{A} + B(\alpha))^{-1}$, as well as estimate (3.12), follows from the well-known inverse operator theorem. Next, we employ the operator identity $C^{-1} = D^{-1} - D^{-1}(I + (C - D)D^{-1})^{-1}(C - D)D^{-1}$, where $C = (\tilde{A} + B(\alpha))$ and $D = A + B(\alpha)$, and based on inequalities (3.14) and (3.15), we get estimate (3.13). $\square$

Theorem 3.1.6. *Let the conditions in Lemma 3.1.5 be fulfilled, i.e. the parameter α is coordinated with noise level δ. Then, RE (3.6) has a unique solution, $\tilde{x}_\alpha$. Moreover, if A is Fredholm operator and x^* is B-normal solution of equation (3.1), $B(0)x^* = 0$, then the following estimate is fulfilled:*

$$\|\tilde{x}_\alpha - x^*\| \leq S(\alpha, x^*) + \frac{\delta c(|\alpha|)}{1 - q}\left(1 + \|x^*\| + S(\alpha, x^*)\right) \tag{3.16}$$

and $\{\tilde{x}_\alpha\}$ converges to x^* at a rate determined by bound (3.16) as $\delta \to 0$.

Proof. The existence and uniqueness of the sequence $\{\tilde{x}_\alpha\}$ as a solution of RE (3.6) for $\alpha \in S$ are proved in Lemma 3.1.5. Since $(\tilde{A} + B(\alpha))(\tilde{x}_\alpha - x^*) = \tilde{f} - f - (\tilde{A} - A)x^* - B(\alpha)x^*$, we get the desired bound (3.16), $\|\tilde{x}_\alpha - x^*\| \leq \|(\tilde{A} + B(\alpha))^{-1}\|(\|\tilde{f} - f\| + \|(\tilde{A} - A)x^*\| + \|(\tilde{A} + B(\alpha))^{-1}B(\alpha)x^*\|) \leq S(\alpha, x^*) + \frac{\delta c(|\alpha|)}{1-q}(1 + \|x^*\| + S(\alpha, x^*))$, based on the proved estimates (3.12), (3.13), and (3.8). Since x^* is a B-normal solution, $\lim_{\alpha \to 0} S(\alpha, x^*) = 0$.

And thanks to the parameter α coordinated with noise level δ, we have $\lim_{\delta \to 0} \delta c(|\alpha|) = 0$. Hence, due to bound (3.16), $\lim_{\delta \to 0} \|\tilde{x}_\alpha - x^*\| = 0$ which completes the proof. $\qquad \square$

As a footnote to the section, it is to be noted that for practical applications of this theorem, one needs algorithms on the choice of SO $B(\alpha)$ and the B-normal solution existence conditions. It's also useful to know the necessary and sufficient conditions for the existence of B-normal solutions x^* to the exact equation (3.1). We discuss these issues as follows.

3.1.3 *Stabilising operator $B(\alpha)$ selection, B-normal solution, existence and correctness class*

If A is a Fredholm operator, $\{\phi_i\}_1^n$ is a basis in $\mathcal{N}(A)$, and $\{\psi_i\}_1^n$ is a basis in $\mathcal{N}(A^*)$, then (referring to Trenogin, 1980, Section 22), one may assume that $B(\alpha) \equiv \sum_{i=1}^n \langle \cdot, \gamma_i \rangle z_i$, where $\{\gamma_i\}$, $\{z_i\}$ are selected such that

$$\det[\langle \phi_i, \gamma_k \rangle]_{i,k=1}^n \neq 0, \quad \langle z_i, \psi_k \rangle = \begin{cases} 1 & \text{if } i = k \\ 0 & \text{if } i \neq k. \end{cases}$$

The equation

$$Ax = f - \sum_{i=1}^n \langle f, \psi_i \rangle z_i \tag{3.17}$$

is resolvable for an arbitrary source function f. Let us now recall $\tilde{f}$, which is the δ-approximation of f. Then, the perturbed equation $Ax + \sum_{i=1}^n \langle x, \gamma_i \rangle z_i = \tilde{f} - \sum_{i=1}^n \langle \tilde{f}, \psi_i \rangle z_i$ has a unique solution $\tilde{x}$ such that $\|\tilde{x} - x^*\| \to 0$ for $\delta \to 0$, where x^* is a unique solution of exact equation (3.17) for which $\langle x^*, \gamma_i \rangle = 0$, $i = \overline{1, n}$. Thus, in the case of the Fredholm operator A as a stabilising operator, one can consider the finite-dimensional operator $B = \sum_1^n \langle \cdot, \gamma_i \rangle z_i$ which does not depend on the parameter α. Of course, with this choice of the SO B, it is necessary to have information about the kernel of the operator A and its defect subspace. Therefore, it is of interest to provide recommendations on the choice of the SO $B(\alpha)$ without relying on such information. It is important to consider a more complex problem of solving equations of the first kind, in which the range of the operator A is not closed. In the following, we consider the generalisation of such results when $B = B(\alpha)$, $\alpha \in S \subset \mathbb{R}^n$.

Theorem 3.1.7. *Let $\|(A + B(\alpha))^{-1}\| \leq c(|\alpha|)$, $\|B(\alpha)\| \leq d(|\alpha|)$ for $\alpha \in S \subset \mathbb{R}^n$, where $c(|\alpha|)$, $d(|\alpha|)$ are continuous functions, with $\lim_{|\alpha| \to 0} c(|\alpha|) = \infty$, $\lim_{|\alpha| \to 0} d(|\alpha|) = 0$. Let $\lim_{|\alpha| \to 0} c(|\alpha|)d(|\alpha|) < \infty$,*

$\mathcal{N}(A) = 0$, and $\overline{R(A)} = Y$. Then, the unique solution x^* of equation (3.1) is a B-normal solution and the operator $B(\alpha)$ is its SO.

Proof. First, let $B(\alpha)x^* \in \mathbb{R}(A)$ for $\alpha \in S$. Then, there exists an element $x_1(\alpha)$ such that $Ax_1(\alpha) = B(\alpha)x^*$. Then, $(A + B(\alpha))^{-1}B(\alpha)x^* = (A + B(\alpha))^{-1}(Ax_1(\alpha) + B(\alpha)x_1(\alpha) - B(\alpha)x_1(\alpha)) = x_1(\alpha) - (A + B(\alpha))^{-1}B(\alpha)x_1(\alpha)$. Since $B(0) = 0$ and $\mathcal{N}(A) = \{0\}$, $\lim_{S \ni \alpha \to 0} x_1(\alpha) = 0$. It is to be noted that by condition, $\|(A + B(\alpha))^{-1}B(\alpha)\| \leq c(|\alpha|)d(|\alpha|)$, where $c(|\alpha|)d(|\alpha|)$ is a continuous function such that the limit $\lim_{|\alpha| \to 0} c(|\alpha|)d(|\alpha|)$ is finite. Then, the α-sequence $\{\|(A + B(\alpha))^{-1}B(\alpha)x^*\|\}$ is infinitesimal when $S \ni \alpha \to 0$. The sequence of operators $\{(A + B(\alpha))^{-1}B(\alpha)\}$ converges pointwise to the zero operator on the linear manifold $L_0 = \{x \mid B(\alpha)x \in R(A)\}$. Thus, we have proved that the theorem is true when $B(\alpha)x^* \in R(A)$. Since by condition $\sup_{\alpha \in S} c(|\alpha|)d(|\alpha|) < \infty$, the α-sequence $\{\|(A + B(\alpha))^{-1}B(\alpha)\|\}$ is bounded. Therefore, the sequence of linear operators $\{(A + B(\alpha))^{-1}B(\alpha)\}$ in space X converges pointwise to the zero operator on the linear manifold $L_0 = \{x \mid B(\alpha)x \in R(A)\}$. But then, on the basis of the Banach–Steinhaus theorem, we have pointwise convergence of this operator's sequence to the zero operator on the closure $\overline{L_0}$, i.e. when $B(\alpha)x^* \in \overline{R(A)}$. Since $\overline{R(A)} = Y$ and $B(\alpha) \in \mathcal{L}(X \to Y)$, $B(\alpha)x^* \in Y$, and Theorem 3.1.7 is proved. $\qquad \square$

The conditions of Theorem 3.1.7 can be relaxed, as noted in the following.

Corollary 3.1.8. *If* $\overline{R(A)} \subset Y$, $\lim_{S \ni \alpha \to 0} x_1(\alpha) = 0$, *then the solution* x^* *of exact equation (3.1) is B-normal iff $B(\alpha)x^* \in \overline{R(A)}$.*

It is to be noted that in Theorem 3.1.7, we used the assumption on the finite limit that $\lim_{S \ni \alpha \to 0} \|(A + B(\alpha))^{-1}\| \|B(\alpha)\|$. We can relax this limitation as well if we assume $B(\alpha) = \alpha B$.

Theorem 3.1.9. *Let* $\|(A + \alpha B)^{-1}\| \leq c(\alpha)$, *where* $\alpha \in \mathbb{R}^1$ *and* $c(\alpha) : (0, \alpha_0] \to \mathbb{R}^+$ *is a continuous function. Suppose that there is a positive integer $n \geq 1$ such that $\lim_{\alpha \to 0} c(\alpha)\alpha^i = \infty$, $i = \overline{0, n - 1}$, and $\lim_{\alpha \to 0} c(\alpha)\alpha^n < \infty$. Let x_0 satisfy equation (3.1), and in the case of $n \geq 2$, there exist $x_1, \ldots, x_{n-1}$ which satisfy the sequence of equations $Ax_i = Bx_{i-1}$, $i = 1, \ldots, n-1$. Then, x_0 is a B-normal solution to equation (3.1) iff $Bx_{n-1} \in \overline{R(A)}$.*

Proof. Since $Ax_i = Bx_{i-1}$, we have an equality $(A + \alpha B)^{-1}\alpha Bx_0 = \alpha(A + \alpha B)^{-1}(Ax_1 + \alpha Bx_1 - \alpha Bx_1) = \alpha x_1 - \alpha^2(A + \alpha B)^{-1}Bx_1 = \cdots = \alpha x_1 - \alpha^2 x_2 + \cdots - (-1)^n \alpha^n (A + \alpha B)^{-1}Bx_{n-1}$, where if $\alpha \to 0$, then the first $n - 2$ terms located on the right-hand side are infinitesimal. Using the Banach–Steinhaus theorem, let's ensure that $\{\alpha^n \|(A + \alpha B)^{-1}Bx_{n-1}\|\}$ is infinitesimal. Indeed, if $Bx_{n-1} \in R(A)$, then there exist x_n such that $Ax_n = Bx_{n-1}$. However, in this case, $\alpha^n(A + \alpha B)^{-1}Bx_{n-1} = \alpha^n(A + \alpha B)^{-1}(A + \alpha B - \alpha B)x_n = \alpha^n x_n - \alpha^{n+1}(A + \alpha B)^1 Bx_n$, where $\alpha^{n+1}\|(A + \alpha B)^1 Bx_n\| \le \alpha^{n+1}c(\alpha)\|Bx_n\|$, $\lim_{\alpha \to 0} \alpha^{n+1}c(\alpha) = 0$. Consequently, $\{\|\alpha^n(A+\alpha B)^1 Bx_{n-1}\|\}$ is infinitesimal, and the sequence of linear operators $\{\alpha^n(A+\alpha B)^1 B\}$ pointwise converges to the zero operator on the linear manifold $L = \{x | Bx \in R(A)\}$, and the sequence $\{\|\alpha^n(A+\alpha B)^1 B\|\}$ is bounded. Since $I = \{x \mid Bx \in \overline{R(A)}\}$, we complete the proof by referring to the Banach–Steinhaus theorem. □

Next, we apply Theorem 3.1.7 for the construction of a stable differentiation algorithm.

3.1.4 *On differentiation regularisation*

Let $y : I \subset \mathbb{R} \to \mathbb{R}$ be a continuous and differentiable function on the interval I, and let its derivative $y'(t)$ be continuous on the interval $(a, b) \subset I$. Then, $y(t) - y(+a) - y'(+a)(t - a) = o(t - a)$ as $t \to +a$. Let $\tilde{y} : [a, b] \to \mathbb{R}$ be a bounded function and c, d be values such that $\sup_{a<t<b} |\tilde{f}(t) - f(t)| = \mathcal{O}(\delta)$, where $f(t) = y(t) - y(+a) - y'(+a)(t-a)$ and $\tilde{f} = \tilde{y}(t) - c - d(t - a)$. In applications, the values of $\tilde{y}(t_i)$ are usually known for $t_i = ih \in [a, b]$ such that $|y(t_i) - \tilde{y}(t_i)| = \mathcal{O}(\delta)$. Our objective here is to find $\tilde{y}_i'(t_i)$ with an accuracy of up to ε. Let us introduce the following equations:

$$\int_a^t x(s)\, ds = f(t), \tag{3.18}$$

$$\int_a^t \tilde{x}_\alpha(s)\, ds + \alpha \tilde{x}_\alpha(t) = \tilde{f}(t).$$

Therefore, we have here $A := \int_a^t [\cdot]\, ds$, $R(A) = \left\{ f(t) \in C^{(1)}_{[a,b]}, f(+a) = 0 \right\}$ $=: \overset{\circ}{C}{}^{(1)}_{[a,b]},$, $\overline{R(A)} = \overset{\circ}{C}_{[a,b]}$, $B(\alpha) := \alpha I$, and $X = Y = C_{[a,b]}$. We construct the inverse operator $(A + \alpha I)^{-1} \in \mathcal{L}(C_{[a,b]} \to C_{[a,b]})$ explicitly as $(A + \alpha I)^{-1} = \frac{1}{\alpha} - \frac{1}{\alpha^2} \int_a^t e^{-\frac{t-s}{\alpha}} [\cdot] \, ds$. Since $f(t) \in R(A)$, equation (3.18) has a unique solution: $x^*(t) = A^{-1}f = y'(t) - y'(+a)$. It is to be noted that

$$\|(A + \alpha I)^{-1}\|_{\mathcal{L}(C_{[a,b]} \to C_{[a,b]})} \leq \frac{1}{\alpha} \left(1 + \frac{1}{\alpha} \max_{a \leq t \leq b} \int_a^t e^{-\frac{t-s}{\alpha}} \, ds\right)$$

$$= \frac{1}{\alpha}(2 - e^{-\frac{a-b}{\alpha}}) < \frac{2}{\alpha}.$$

The parameter α should be related with δ, e.g. $\alpha = \sqrt{\delta}$. Since $\|B(\alpha)\| = \alpha$, $c(\alpha) = \frac{2}{\alpha}$, $\mathcal{N}(A) = \{0\}$, based on Theorem 3.1.7, the continuous function $x^*(t) = y'(t) - y'(+a)$ is a B-normal solution iff $x^*(t) \in \overline{R(A)}$. In our case, $R(A) = \overset{\circ}{C}{}^{(1)}_{[a,b]}$. Taking into account the fact that the linear functions space $\overset{\circ}{C}{}^{(1)}_{[a,b]}$, is dense $L_1 = \{f(t) \in C_{[a,b]}, f(+a) = 0\}$, $x^*(+a) = 0$, then $x^*(t) \in \overline{R(A)}$. Therefore, based on the main theorem and Theorem 3.1.7, the formula

$$\tilde{x}_\alpha(t) = \frac{\tilde{y}(t) - c - d(t - a)}{\alpha} - \frac{1}{\alpha^2} \int_a^t e^{-\frac{t-s}{\alpha}} (\tilde{y}(s) - c - d(s - a)) \, ds$$

$$(3.19)$$

defines an algorithm for stable differentiation, $\tilde{y}'(t)$. More precisely, $\forall \varepsilon > 0$ $\exists \delta_0 = \delta_0(\varepsilon) > 0$ such that, if $\sup_{a < t < b} |\tilde{f}(t) - f(t)| \leq \delta, \delta \leq \delta_0(\varepsilon)$, then $\max_{a \leq t \leq b} |\tilde{x}_\alpha(t) - f'(t)| \leq \varepsilon$. If we select $\alpha = \sqrt{\delta}$, then $\lim_{\delta \to 0} \max_{a \leq t \leq b} |\tilde{x}_\alpha(t) - (y'(t) - y'(+a))| = 0$. Therefore, $\{\tilde{x}_\alpha\}$ converges uniformly to $y'(t) - y'(+a)$ as $\delta \to 0$. Based on (3.19), we have constructed a regularised differentiation algorithm which is uniform wrt $t \in [a, b]$. Let us demonstrate its efficiency using an example, as follows.

Consider that we add noise to exact data as $\tilde{y}(t) = y(t) + \delta R(t)$, with noise levels $\delta = 0.1$, $\delta = 0.01$, and $\delta = 0.001$, where $R(t)$ is a random function with a mean value of zero and standard deviation $\sigma = 1$. The number of grid points used is 512. The trapezoidal quadrature rule is used.

For this example, we use the function $y(t) = \frac{1}{t^3+1} \sin\left(\frac{\pi t}{4}\right)$, $t \in [0, 3]$, with its derivative $y'(t) = \frac{-3t^2}{(t^3+1)^2} \sin\left(\frac{\pi t}{4}\right) + \frac{\pi}{4(t^3+1)} \cos\left(\frac{\pi t}{4}\right)$. Figure 3.1(a) and (b) show the exact and computed derivatives and the errors, for the noise level of $\delta = 0.001$.

(a)

(b)

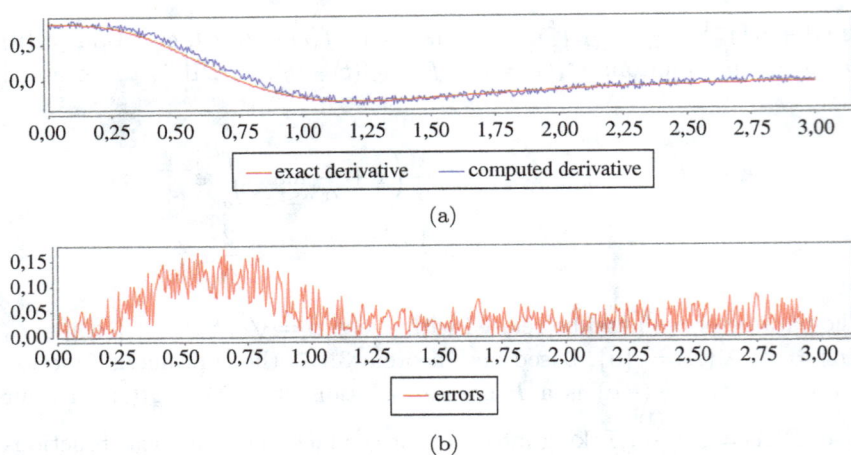

Figure 3.1.   (a) The exact and the computed derivatives $\tilde{y}'(t)$ and (b) the errors. The noise level $\delta = 0.001$ is used to generate $\tilde{y}(t)$.

3.2   Lavrentiev Regularisation of Integral Equations of the First Kind in the Space of Continuous Functions

Lavrentiev regularisation is a classical and powerful method for addressing ill-posed problems in various areas of science, engineering, and mathematics. Ill-posed problems are those for which small changes in the input data can lead to large changes in the output, making traditional methods of solving them unreliable or ineffective. Ill-posedness often arises in the context of inverse problems, where the data are noisy, incomplete, or subject to uncertainty. Lavrentiev regularisation provides a systematic and effective way to stabilise and regularise these problems, ensuring that valid and useful solutions can be obtained.

Lavrentiev regularisation is essential for a wide range of practical applications in many fields, such as image and signal processing, inverse problems in geophysics, and medical imaging. It offers a systematic framework for addressing and solving problems where traditional methods fail, thereby providing a valuable tool for researchers and practitioners working on real-world engineering problems.

Let us introduce the integral operator $Kx := \int_0^t K(t,s)x(s)\,ds$. We employ the standard notation and conditions:

(A)   *The kernel $K(t,s)$ is defined, continuous, and differentiable with respect to t in the set $D := \{0 < s < t < T\}$, $\min_{0 \le t \le T} |K(t,t)| = d > 0$.*

The set $\mathcal{R}(K) = \overset{\circ}{\mathcal{C}}'_{[0,T]}$ is the range of values of this operator, and the set $\overset{\circ}{\mathcal{C}}_{[0,T]}$ is based on the Weierstrass approximation theorem, forming the closure region $\mathcal{R}(K)$.

(B) *Let $f : (0,T) \to \mathbb{R}^1$ be a differentiable function, $f(0) = 0$, and let α be a small positive parameter.*

Let us introduce the integral equation

$$Kx = f, \tag{3.20}$$

the perturbed equation

$$Kx + \alpha x = f, \tag{3.21}$$

and the approximate equation

$$\widetilde{K}x + \alpha x = \tilde{f}. \tag{3.22}$$

The objective here is to find the continuous solution. In connection with the problem of numerically solving the Volterra equations, let's understand how the solutions of equations (3.20), (3.21), and (3.22) are linked. There is an extensive literature on this issue and the problem of regularisation, an important part of which consists of the results of Lavrentiev and Denisov, which served as an impetus for this section.

A feature of the methodology in this section is the systematic use of the concepts of stabilising operators, estimates and the norm of the inverse operator

$$\|(K + \alpha I)^{-1}\|_{\mathcal{L}(\mathcal{C}_{[0,T]} \to \mathcal{C}_{[0,T]})}$$

in operator topology and some results related to the classical theory of perturbation of linear operators and the Banach–Steinhaus theorem (Trenogin, 1980). In this way, it was possible to strengthen A. M. Denisov's (1975) result on the regularisation of the Volterra equation. As a result, such restrictions have been lifted, namely the assumption of the exact specification of the kernel $K(t, s)$, the existence of second derivatives with respect to kernel t and free function in equation (3.20).

If conditions (A) and (B) are met, equation (3.20) is fair and equivalent to the integral Volterra equation of the second kind:

$$K(t,t)x(t) + \int_0^t K_t'(t,s)x(s)\,ds = f'(t). \tag{3.23}$$

Lemma 3.2.1. *Let conditions (A) and (B) be satisfied. Then:*

(1) *The linear operator $K \in \mathcal{L}(\mathcal{C}_{[0,T]} \to \overset{\circ}{\mathcal{C}}'_{[0,T]},)$ has a bounded inverse, and equation (3.20) has a unique solution, $x^*(t)$, in the class $\mathcal{C}_{[0,T]}$:*
$x^*(0) = \frac{f'(0)}{K(0,0)}.$

(2) *If, in addition to conditions (A) and (B), the function $f(t)$ is twice differentiable and the kernel $K(t,s)$ is differentiable with respect to s and twice with respect to t, then the solution $x^*(t) \in \mathcal{C}'_{[0,T]}$. If $f'(0) = 0$, then $x^*(t) \in \mathcal{R}(K)$, and the system*

$$\begin{cases} Kx_1 = f, \\ Kx_2 = x_1 \end{cases} \tag{3.24}$$

enjoys the unique solution $x_1(t) \in \overset{\circ}{\mathcal{C}}'_{[0,T]},$, $x_2(t) \in \mathcal{C}_{[0,T]}$.

Proof. Let's check the validity of statement (1):

(a) $Kx\big|_{t=0} = \int_0^t K(t,s)x(s)\,ds\big|_{t=0} = 0,$

(b) $\frac{d}{dt}Kx = K(t,t)x(t) + \int_0^t K'_t(t,s)x(s)\,ds$ — a continuous function for any function $x(t) \in \mathcal{C}_{[0,T]}$.

From (a) and (b), it follows that K is a linear bounded operator acting from $\mathcal{C}_{[0,T]}$ on $\overset{\circ}{\mathcal{C}}'_{[0,T]}$. That is, the inclusion of $K \in \mathcal{L}(\mathcal{C}_{[0,T]} \to \overset{\circ}{\mathcal{C}}'_{[0,T]},)$ is proved. Because equation (3.23) is equivalent to equation (3.20), the inverse operator K^{-1} is constructed explicitly according to the formula

$$x = K^{-1}f = \frac{f'(t)}{K(t,t)} + \int_0^t R(t,s)\frac{f'(s)}{K(s,s)}\,ds,$$

where $R(t,s)$ is the resolvent of the kernel $-\frac{K'_t(t,s)}{K(t,t)}$. The solution $x^*(t)$ is unique, $x^*(0) = \frac{f'(0)}{K(0,0)}$. If for $0 \le s \le t \le T$ the estimates $|K'_t(t,s)| \le C$ are satisfied, with $|K(t,t)| \ge d > 0$, then (3.23) implies the inequality

$$|x(t)| \le \frac{1}{d}\|f\|_{\overset{\circ}{\mathcal{C}}'_{[0,T]},} + \int_0^t \frac{C}{d}|x(s)|\,ds.$$

That's why,

$$|x(t)| \le \frac{1}{d}e^{\frac{C}{d}t}\|f\|_{\overset{\circ}{\mathcal{C}}'_{[0,T]}}.$$

Hence,

$$||K^{-1}||_{\mathcal{L}(\overset{\circ}{\mathcal{C}}'_{[0,T]}\,\to\,\mathcal{C}_{[0,T]})} \leq \frac{1}{d}e^{\frac{\mathcal{S}}{d}T}.$$

The validity of statement (1) has been established. Let's prove that statement (2) is fair. From equation (3.23), equivalent to equation (3.20), it follows that if the conditions formulated in (2) are met, the solution $x^* \in \mathcal{C}'_{[0,T]}$. If $f'(0) = 0$, then $x^* \in \overset{\circ}{\mathcal{C}}'_{[0,T]} = \mathcal{R}(K)$, and the system (3.24) will be solvable, which is what we needed to prove. $\qquad\square$

Remark 3.2.2. If $\min_{0\leq t\leq T}|K(t,t)| = 0$, then equation (3.20) may turn out to be unsolvable or have a non-unique solution. For example, the equation $\int_0^t (t - 2s)x(s)\,ds = t^2$, where $K(t,t) = -t$ satisfies the family $x(t,c) = c - 2t$, where c is a constant, and the equation $\int_0^t tx(s)\,ds = \sin t$ is not solvable in the class $\mathcal{C}_{[0,T]}$.

3.2.1 Solutions of original and regularised equations

Let us introduce the α-family of operators $B_\alpha := \alpha(K+\alpha I)^{-1}$ corresponding to equation (3.23), considered for $\alpha > 0$ and acting from $\mathcal{C}_{[0,T]}$ on $\mathcal{C}_{[0,T]}$. The operator B_α satisfies the following two properties:

Property 1. *If the sequence $||B_\alpha u||$ for $\alpha \to +0$ is infinitesimal, then $u \in \overline{\mathcal{R}(K)}$. Indeed, let us write down the obvious operator identity for $\alpha > 0$:*

$$K(K + \alpha I)^{-1} + B_\alpha = I.$$

For any u and nonnegative α, this identity implies the equality

$$K(K + \alpha I)^{-1}u + B_\alpha u = u.$$

Since by condition the numerical sequence $\{||B_\alpha u||\}$ is infinitesimal as $\alpha \to +0$, the sequence $\{||u - Ku_\alpha||\}$, where $u_\alpha = (K + \alpha I)^{-1}u$, should also be infinitesimal for $\alpha \to +0$. However, it is necessary that the inclusion $u \in \overline{\mathcal{R}(K)}$ be satisfied. Thus, the validity of Property 1 is established.

Property 2. *Let x^* be a solution to equation (3.20) and x_α be a solution to equation (3.21). Let the sequence $\{||B_\alpha x^*||\}$ be infinitesimal for $\alpha \to 0$. Then, $\lim_{\alpha\to+0}||x_\alpha - x^*|| = 0$.*

The proof follows from the identity $x^* - x_\alpha = \alpha(K + \alpha I)^{-1}x^*$ because $B_\alpha = \alpha(K + \alpha I)^{-1}$.

Lemma 3.2.3. *Let condition (A) be satisfied and $K(t,t) = 1$. Then, for $\alpha > 0$, the following estimate holds: $\|B_\alpha\|_{\mathcal{L}(C \to C_{0,T})} \leq 2e^{aT}$, where $a = \max_{0 \leq t \leq T} |K'_t(t,s)|$.*

Proof. Introducing the notation $K_0 x := \int_0^t x(s)\,ds$, $(K - K_0)x := \int_0^t (K(t,s) - 1)x(s)\,ds$, we rewrite equation (3.22) in the form $\alpha x + K_0 x + (K - K_0)x = f$.

Note that for $\alpha \neq 0$, the bounded operator $\alpha I + K_0$ acts one-to-one from $\mathcal{C}_{[0,T]}$ to $\mathcal{C}_{[0,T]}$, and its inverse is limited and has the form

$$(\alpha I + K_0)^{-1} = \frac{1}{\alpha} I - \frac{1}{\alpha^2} \int_0^t e^{-\frac{t-z}{\alpha}} [\cdot]\,dz.$$

Therefore, equation (3.22) is reduced by multiplication with the bounded operator $(\alpha I + K_0)^{-1}$ to the equivalent equation

$$x + \frac{1}{\alpha}(K - K_0)x - \frac{1}{\alpha^2} \int_0^t e^{-\frac{t-z}{\alpha}} (K - K_0)x\big|_z \, dz$$

$$= \frac{1}{\alpha} f(t) - \frac{1}{\alpha^2} \int_0^t e^{-\frac{t-z}{\alpha}} f(z)\,dz. \tag{3.25}$$

Here and further, $(K - K_0)x\big|_z := \int_0^z (K(z,s) - 1)x(s)\,ds$. Changing the order of integration, we obtain the equality

$$\int_0^t e^{-\frac{t-z}{\alpha}} (K - K_0)x\big|_z \, dz = \int_0^t e^{-\frac{t-z}{\alpha}} \int_0^z (K(z,s) - 1)x(s)\,ds\,dz$$

$$= \int_0^t \int_s^t e^{-\frac{t-z}{\alpha}} (K(z,s) - 1)\,dz\,x(s)\,ds.$$

Integrating by parts, taking into account the identity $K(t,t) \equiv 1$, we arrive at the equality

$$\int_s^t e^{-\frac{t-z}{\alpha}} (K(z,s) - 1)\,dz$$

$$= e^{-\frac{t}{\alpha}} \left\{ \alpha e^{z/\alpha}(K(z,s) - 1)\Big|_{z=s}^{z=t} - \alpha \int_s^t e^{z/\alpha} K'_z(z,s)\,dz \right\}$$

$$= \alpha(K(t,s) - 1) - \alpha \int_s^t e^{-\frac{t-z}{\alpha}} K'_z(z,s)\,dz.$$

Let us introduce the function $Q(t,s,\alpha) := \int_s^t e^{-\frac{t-z}{\alpha}} K'_z(z,s)\,dz$.

Taking this notation into account, we have the identity

$$\frac{1}{\alpha^2} \int_0^t e^{-\frac{t-z}{\alpha}} (K - K_0)x\Big|_z \, dz$$

$$= \frac{1}{\alpha} \int_0^t (K(t,s) - 1)x(s) \, ds - \int_0^t \frac{Q(t,s,\alpha)}{\alpha} x(s) \, ds,$$

due to which equation (3.25) takes the form

$$x(t) + \int_0^t \frac{Q(t,s,\alpha)}{\alpha} x(s) \, ds = \frac{1}{\alpha} f(t) - \frac{1}{\alpha^2} \int_0^t e^{-\frac{t-z}{\alpha}} f(z) \, dz. \qquad (3.26)$$

Let us show that the kernel $\frac{Q(t,s,\alpha)}{\alpha}$ is continuous in the domain $D = \{0 < s < t < T, 0 < \alpha < \infty\}$ and is bounded on the compact set $\overline{D}$. Note that

$$|Q(t,s,\alpha)| \le \int_s^t e^{-\frac{t-z}{\alpha}} |K'_z(z,s)| \, dz,$$

where $|K'_z(z,s)| \le a$. Therefore,

$$|Q(t,s,\alpha)| \le ae^{-\frac{t}{\alpha}} \int_s^t e^{z/\alpha} \, dz = ae^{-t/\alpha} \alpha(e^{t/\alpha} - e^{s/\alpha}) = a\alpha(1 - e^{-\frac{t-s}{\alpha}})$$

$$\le a\alpha(1 - e^{-\frac{T}{\alpha}}) < a\,\alpha.$$

Then, the kernel $\frac{Q(t,s,\alpha)}{\alpha}$ is continuous in the region $D = \{0 < s < t < T, 0 < \alpha < \infty\}$ and has a continuous resolvent, $R(t,s,\alpha)$. Since $\frac{Q(t,s,\alpha)}{\alpha} < a$, the resolvent satisfies the estimate $|R(t,s,\alpha)| < ae^{a(t-s)}$. Using the resolvent $R(t,s,\alpha)$, the solution to equation (3.22) is constructed in closed form using the formula

$$x(t) = \frac{f(t)}{\alpha} - \frac{1}{\alpha^2} \int_0^t e^{-\frac{t-z}{\alpha}} f(z) \, dz + R\left(\frac{f(s)}{\alpha} - \frac{1}{\alpha^2} \int_0^s e^{-\frac{s-z}{\alpha}} f(z) \, dz\right).$$

Here, $R[\cdot] := \int_0^t R(t,s,\alpha)[\cdot] \, ds$. Therefore, the inequality is true:

$$\max_{0 \le t \le T} \left| R\left(\frac{f(s)}{\alpha} - \frac{1}{\alpha^2} \int_0^s e^{-\frac{s-z}{\alpha}} f(z) \, dz\right) \right| \le \frac{2}{\alpha} \|f\| \max_{0 \le t \le T} a \int_0^t e^{a(t-s)} \, ds,$$

where $\max_{0 \le t \le T} a \int_0^t e^{a(t-s)} \, ds = e^{aT} - 1$. Considering the assessment

$$\max_{0 \le t \le T} \left| \frac{1}{\alpha} f(t) - \frac{1}{\alpha^2} \int_0^t e^{-\frac{t-z}{\alpha}} f(z) \, dz \right| \le \frac{2}{\alpha} \|f\|,$$

we arrive at the inequality

$$||x|| = ||(K + \alpha I)^{-1} f|| \leq \frac{2}{\alpha} e^{aT} ||f||$$

for $f(t) \in \mathcal{C}_{[0,T]}$. Therefore, $||(K + \alpha I)^{-1}||_{\mathcal{L}(\mathcal{C}_{[0,T]} \to \mathcal{C}_{[0,T]})} \leq \frac{2}{\alpha} e^{aT}$. Since $B_\alpha = \alpha(K + \alpha I)^{-1}$, $||B_\alpha||_{\mathcal{L}(\mathcal{C}_{[0,T]} \to \mathcal{C}_{[0,T]})} \leq 2 e^{aT}$, which is what we needed to prove. $\qquad \square$

Remark 3.2.4. A similar result is obtained in the case of $\min_{a < t \leq T} |K(t,t)| = d > 0$. Indeed, this case reduces to considering the equation $\int_0^t A(t,s)x(s)\, ds = \frac{f(t)}{K(t,t)}$, where $A(t,s) := K(t,s)/K(t,t)$, $A(t,t) = 1$.

Lemma 3.2.5. *Let the kernel $K(t,s)$ be continuous and differentiable with respect to t, $||B_\alpha||_{\mathcal{L}(\mathcal{C}_{[0,T]} \to \mathcal{C}_{[0,T]})} \leq C$ for $\alpha > 0$. Then, for any function $u(t) \in \overset{\circ}{\mathcal{C}}_{[0,T]}$, the sequence $\{||B_\alpha u||\}$ will be infinitesimal at $\alpha \to +0$.*

Proof. If $u(t) \in \overset{\circ}{\mathcal{C}'}_{[0,T]}$, then $u(t) \in \mathcal{R}(K)$; therefore, in the space $\mathcal{C}_{[0,T]}$, there is a function $x(t)$ such that $Kx = u$. Then, $B_\alpha u = (K + \alpha I)^{-1} \alpha(Kx + \alpha x - \alpha x) = \alpha x - \alpha^2 (K + \alpha I)^{-1} x = \alpha x - \alpha B_\alpha x$. Therefore, for $u(t) \in \overset{\circ}{\mathcal{C}'}_{[0,T]}$, the estimate $||B_\alpha u|| \leq \alpha(1 + C)||x||$ and $\lim_{\alpha \to 0} ||B_\alpha u|| = 0$ hold.

Now, let $u(t) \in \overset{\circ}{\mathcal{C}}_{[0,T]}$, where $\overset{\circ}{\mathcal{C}}_{[0,T]}$ is the closure of the range of values $\mathcal{R}(K)$. By condition, the α-sequence $\{||B_\alpha||\}$ is limited. According to the proven $B_\alpha u \to 0$ for $\forall u \in \overset{\circ}{\mathcal{C}'}_{[0,T]}$. Linear combinations of functions from $\overset{\circ}{\mathcal{C}'}_{[0,T]}$ lie densely everywhere in $\overset{\circ}{\mathcal{C}}_{[0,T]}$. Therefore, based on the Banach–Steinhaus theorem (see Lusternik, 1956, Theorem 4, p. 151) $B_\alpha u \to 0$ for $\forall u \in \overset{\circ}{\mathcal{C}}_{[0,T]}$, which is what we needed to prove. $\qquad \square$

Theorem 3.2.6. *Let $K(t,s)$ and $f(t)$ be continuous and differentiable with respect to t, $f(0) = 0$, $f'(0) = 0$. Then, in the class $\overset{\circ}{\mathcal{C}}_{[0,T]}$, equation (3.20) has a unique solution, $x^*(t)$. In this case, $\lim_{\alpha \to +0} ||x_\alpha - x^*||_{\mathcal{C}_{[0,T]}} = 0$, where x_α is the only continuous solution to equation (3.21).*

Proof. The existence and uniqueness of the solutions of equation (3.20) in the class $\overset{\circ}{\mathcal{C}}_{[0,T]}$ are proved in Lemma 3.2.1. By virtue of Property 2, $x^* - x_\alpha = B_\alpha x^*$, where, based on Lemma 3.2.3, $||B_\alpha||_{\mathcal{L}(\mathcal{C}_{[0,T]} \to \mathcal{C}_{0,T})} \leq C$. Since $x^* \in \overset{\circ}{\mathcal{C}}_{[0,T]}$, based on Lemma 3.2.5, the sequence $||B_\alpha x^*||_{\mathcal{C}_{[0,T]}}$ will be infinitesimal for $\alpha \to +0$. The theorem is proven. $\qquad \square$

3.2.2 The main theorem

Let us introduce the functions $\widetilde{K}(t,s)$, $\widetilde{f}(t)$ defined in the domain D, such that the following inequalities are satisfied: $\sup_{0<s<t<T}|\widetilde{K}(t,s)-K(t,s)| \leq \delta$, $\sup_{0<t<T}|\widetilde{f}(t)-f(t)| \leq \delta$. We assume that the functions $\widetilde{K}(t,s)$, $\widetilde{f}(t)$ in the domain D have a finite number of discontinuity points of the first kind. Let E denote the Banach space of piecewise continuous functions with with a finite number of discontinuity points of the first kind and with norm $||x||_E = \sup_{0<t<T}|x(t)|$. In equation (3.22), we set $\widetilde{K} = \int_0^t \widetilde{K}(t,s)[\cdot]\,ds$, $\widetilde{f} = \widetilde{f}(t)$. Obviously, $\widetilde{K} \in \mathcal{L}(E \to E)$ because $||\widetilde{K}|| \leq T \cdot \sup_{0<s<t<T}|\widetilde{K}(t,s)|$. Therefore, equation (3.22) with the Volterra operator $\widetilde{K}$ in space E has a unique solution $\widetilde{x}_\alpha$.

Let $||(K+\alpha I)^{-1}|| \leq c/\alpha$ for $\alpha > 0$. This estimate is satisfied, for example, under the conditions in Lemma 3.2.3. Further, in equations (3.21) and (3.22), let $\alpha = \delta^\nu$, for $\nu \in (0,1)$. Then, based on the inverse operator theorem (see, for example, Lusternik, 1956, Theorem 2, p. 156), there is a positive δ_0 such that for $\forall \delta \in (0, \delta_0]$, the asymptotic estimate $||(\widetilde{K}+\alpha I)^{-1}||_{\mathcal{C}_{[0,T]} \to \mathcal{C}_{[0,T]}} = \mathcal{O}(1/\alpha)$. Moreover, for any $\varepsilon > 0$, $\exists \delta_0 = \delta_0(\varepsilon) > 0$ such that for $\delta \in (0, \delta_0]$, the inequalities are satisfied:

$$||(\widetilde{K}+\alpha I)^{-1} - (K+\alpha I)^{-1}||_{\mathcal{L}(E\to E)} \leq \varepsilon/3, \tag{3.27}$$

$$\delta||(\widetilde{K}+\alpha I)^{-1}||_{\mathcal{L}(E\to E)} \leq \varepsilon/3. \tag{3.28}$$

Recall that the solution x^* to equation (3.20) is an element in the space $\overset{\circ}{\mathcal{C}}_{[0,T]}$. Therefore, based on Lemma 3.2.5, for any fixed $\varepsilon > 0$, $\exists \delta_1 = \delta_1(\varepsilon) > 0$ is such that for any $\delta \in (0, \delta_1]$, the inequality

$$||(K+\alpha)^{-1}\alpha x^*||_E \leq \varepsilon/3 \tag{3.29}$$

will be satisfied. Let $\delta^* = \min\{\delta_0(\varepsilon), \delta_1(\varepsilon)\}$. Then, for $\delta \in (0, \delta^*]$, inequalities (3.27), (3.28), and (3.29) are simultaneously satisfied.

If $\widetilde{x}_\alpha = (\widetilde{K}+\alpha I)^{-1}\widetilde{f}$ and $x^* = K^{-1}f$ are the corresponding solutions of equations (3.22) and (3.20), then the following inequality will be true:

$$|\widetilde{x}_\alpha - x^*| \leq ||(\widetilde{K}+\alpha I)^{-1}(\widetilde{f}-f)||_E + ||((\widetilde{K}+\alpha I)^{-1} - (K+\alpha I)^{-1})f||_E$$
$$+ ||(K+\alpha I)^{-1}f - K^{-1}f||_E.$$

Since $||(K+\alpha I)^{-1}f - K^{-1}f|| = ||(K+\alpha I)^{-1}\alpha x^*||$, then taking into account estimates (3.27), (3.28), and (3.29) for all $\delta \in (0, \delta^*]$, we have the assessment $||\widetilde{x}_\alpha - x^*||_E \leq \varepsilon$. The main theorem follows from what has been proved.

Theorem 3.2.7. *Let the domain $D = \{0 < s < t < T\}$ be given and the functions $K(t,s)$, $f(t)$ be continuously differentiable with respect to t. Let $f(0) = 0$, $f'(0) = 0$, $\min_{0 \leq t \leq T} |K(t,t)| = d > 0$ and the element $x^*(t)$ from E be a solution to equation (3.20). Let the piecewise continuous functions $\widetilde{K}(t,s)$ and $\tilde{f}(t)$ be defined in the domain D and satisfy the estimates $|\widetilde{K}(t,s) - K(t,s)| \leq \delta$, $|\tilde{f}(t) - f(t)| \leq \delta$. Then, for all $\varepsilon > 0$ and $\alpha = \delta^\nu$, where $\nu \in (0,1)$, $\exists \delta^*(\varepsilon) > 0$ such that for $\delta \in (0, \delta^*(\varepsilon)]$, the only solution to the $\tilde{x}$ equation*

$$\alpha \tilde{x}_\alpha(t) + \int_0^t \widetilde{K}(t,s)\tilde{x}_\alpha(s)\, ds = \tilde{f}(t)$$

satisfies the estimate $\sup_{0 < t < T} |\tilde{x}_\alpha(t) - x^*(t)| \leq \varepsilon$.

Let us consider the examples from the class of equations of the first kind of the form (3.20), having a unique continuous solution, $x : [0,T] \to E_2$, admitting regularisation according to Lavrentiev.

Consider the initial problem

$$\begin{cases} \int_0^t K(t-s)\left(x(s,y) + \frac{\partial x(s,y)}{\partial y}\right)\, ds = f(t,y), \ y \in [a,b], \\ x|_{y=a} = x_0(t), \ t \in [0,T]. \end{cases} \tag{3.30}$$

Let $K(t)$, $f(t,y)$ be continuous functions, differentiable with respect to t, $f(0,y) = 0$, $K(0) = 1$, $E_1 = \mathcal{C}'_{[a,b]}$, and $E_2 = \mathcal{C}_{[a,b]}$. Then, the initial problem (3.30) has a unique continuous solution. Regularisation can be carried out by solving sequentially the equations

$$\alpha u(t,y) + \int_0^t K(t-s)u(s,y)\, ds = f(t,y),$$

$$\frac{\partial x(t,y)}{\partial y} + x(t,y) = u(t,y), \ x|_{y=a} = x_0(t).$$

Let us focus on another example. The integral equation

$$\int_0^t K(t-s)\left[x(s,y) + \int_0^1 Q(y,z)x(s,z)\, dz\right]\, ds = f(t,y), \tag{3.31}$$

where the given functions are continuous and differentiable with respect to t, $f(0,y) = 0$, $K(0) = 1$, -1 is not an eigenvalue of the kernel $Q(y,z)$, and the equation has a unique continuous solution. Regularisation of this example is carried out by sequentially solving the equations

$$\alpha u(t, y) + \int_0^t K(t - s)u(s, y)\, ds = f(t, y),$$

$$x(t, y) + \int_0^1 Q(y, z)x(t, z)\, dz = u(t, y). \tag{3.32}$$

If -1 is an eigenvalue of the kernel $Q(y, z)$, $\phi_1, \ldots, \phi_n$ are its own functions, and $\psi_1, \ldots, \psi_n$ are its own functions of the union core, then based on Sidorov *et al.* (2015), the kernel $Q(y, z)$ in equation (3.31) must be replaced with the perturbed kernel $Q(y, z) + \sum_{i=1}^n \phi_i(y)\psi_i(z)$.

Remark 3.2.8. In Section 5.4, the efficiency of the Lavrentiev regularisation is demonstrated for the problem of energy storage Volterra model regularisation.

3.2.3 *Integral equations with discontinuous kernels*

Let us consider the use of the perturbation method in the construction of regularised numerical methods for solving Volterra integral equations of the first kind of the form

$$\int_0^t K(t, s)x(s)ds = f(t), \quad 0 \le s \le t \le T, \; f(0) = 0. \tag{3.33}$$

The kernel $K(t, s)$, defined in the region $\{0 \le s \le t \le T\}$, undergoes discontinuities of the first kind on the curves $s = \alpha_i(t)$ and is given by the formula

$$K(t, s) = \begin{cases} K_1(t, s), \; t, s \in m_1, & m_i = \{t, s \mid \alpha_{i-1}(t) < s < \alpha_i(t)\}, \\ \cdots \quad \cdots\cdots\cdots \\ K_n(t, s), \; t, s \in m_n, & \alpha_0(t) = 0, \; \alpha_n(t) = t, \; i = \overline{1, n}. \end{cases} \tag{3.34}$$

Here, the functions $\alpha_i(t)$, $f(t)$, and $K_i(t, s)$ are continuous in their respective domains and admit continuous extensions to compact sets $\overline{m_i}$, $K_n(t, t) \ne 0$, $\alpha_i(0) = 0$, and $0 < \alpha_1(t) < \alpha_2(t) < \cdots < \alpha_{n-1}(t) < t$, for $0 < t \le T$. $\alpha_i(t)$ increase in a small neighbourhood $0 \le t \le \tau$, with $0 \le \alpha_1'(0) \le \cdots \le \alpha_{n-1}'(0) < 1$. An important role in the theory of such equations, constructed in monographs by Sidorov (2013a, 2014b), is played by the function $D(t) = \sum_{i=1}^{n-1} \left| \alpha_i'(t)K_n^{-1}(t, t) \right| \cdot \left| K_i(t, \alpha_i(t)) - K_{i+1}(t, \alpha_i(t)) \right|$ and the characteristic exponential equation

$$K_n(0,0) + \sum_{i=0}^{n-1} (\alpha_i'(0))^{1+j} (K_i(0,0) - K_{i+1}(0,0)) = 0.$$

If $D(0) < 1$, then the characteristic equation has no natural roots, and equation (3.33) (referring to the monograph by Sidorov (2013a)) has a unique solution in the class of continuous functions $C_{[0,T]}$. Let us present the results of a numerical solution to an equation with discontinuous kernels that satisfy the conditions formulated above.

We demonstrate the efficiency of the proposed method using the following example. Consider a linear equation of the first kind of the form (3.33):

$$\int_0^{t/4} (1+t+s)x(s)\,ds + \int_{t/4}^{t/2} (2+ts)x(s)\,ds + \int_{t/2}^{t} (1+t+s)x(s)\,ds$$

$$= \frac{31t^6}{40960} + \frac{1099t^5}{20480} + \frac{271t^4}{8192}, \quad t \in [0,2].$$

Here, the exact solution is $\bar{x}(t) = t^3/8$. Let us present the results of the algorithm using the Lavrentiev regularisation described above for the case when the search for an approximate solution is performed in the form of a piecewise constant (PWC) function:

$$x_N(t) = \sum_{i=1}^{N} x_i \delta_i(t), \ t \in (0,T], \ \delta_i(t) = \begin{cases} 1, & \text{for } t \in \Delta_i = (t_{i-1}, t_i]; \\ 0, & \text{for } t \notin \Delta_i. \end{cases}$$

A detailed description of the algorithm is given in the work of Sidorov *et al.* (2014). We use a uniform grid here.

Table 3.1 shows the errors $\varepsilon_h = \max_{0 \le i \le N} |x_N(t_i) - \bar{x}(t_i)|$, obtained by matching the step with the levels of normally distributed random noise added to the kernel and to the right-hand side. Calculations for non-noisy data are also shown for comparison. The values of the α parameter are chosen in such a way as to ensure the minimum error at a given step. Also, Table 3.1 shows the results of calculations using the piecewise linear (PWL) approximation of the solution:

$$x_N(t) = \sum_{i=1}^{N} \left(x_{i-1} + \frac{x_i - x_{i-1}}{t_i - t_{i-1}} (t - t_{i-1}) \right) \delta_i(t), \ t \in (0,T],$$

where $\delta_i(t)$ is defined as above.

Table 3.1. Numerical analysis of Gauss and midpoint quadrature rules.

			10^{-1}	10^{-2}	10^{-3}	10^{-4}
		h				
			Without noise			
No reg.	Midpoint rect.	PWC	0.139529	0.014874	0.001497	0.000150
		PWL	$5.39 \cdot 10^{-3}$	$2.43 \cdot 10^{-4}$	$2.1 \cdot 10^{-5}$	$2.0 \cdot 10^{-6}$
	Gauss	PWC	0.139529	0.014874	0.001497	0.000150
		PWL	0.004932	$5.0 \cdot 10^{-5}$	$5.0 \cdot 10^{-7}$	$6.0 \cdot 10^{-9}$
With regularisation	Midpoint rect.	PWC	0.139529	0.014874	0.001497	0.000150
		α	10^{-4}	10^{-6}	10^{-8}	10^{-10}
		PWL	$5.39 \cdot 10^{-3}$	$2.43 \cdot 10^{-4}$	$1.1 \cdot 10^{-5}$	$3.7 \cdot 10^{-7}$
		α	10^{-1}	10^{-2}	10^{-3}	10^{-4}
	Gauss	PWC	0.139529	0.014874	0.001497	0.000150
		α	10^{-4}	10^{-6}	10^{-8}	10^{-10}
		PWL	0.004932	$5.0 \cdot 10^{-5}$	$5.0 \cdot 10^{-7}$	$5.5 \cdot 10^{-9}$
		α	10^{-1}	10^{-2}	10^{-3}	10^{-4}
		With noise δ	10^{-2}	10^{-4}	10^{-6}	10^{-8}
No reg.	Midpoint rect.	PWC	0.137164	0.015311	0.001596	0.000165
		PWL	0.261607	0.079015	0.023068	0.006592
	Gauss	PWC	0.133109	0.015482	0.001624	0.000165
		PWL	0.098011	0.076760	0.024611	0.008765
With regularisation	Midpoint rect.	PWC	0.124463	0.014720	0.001568	0.000162
		α	10^{-4}	10^{-6}	10^{-8}	10^{-10}
		PWL	0.044884	0.006480	0.000850	0.000096
		α	10^{-1}	10^{-2}	10^{-3}	10^{-4}
	Gauss	PWC	0.123097	0.014767	0.001551	0.000162
		α	10^{-2}	10^{-4}	10^{-6}	10^{-8}
		PWL	0.037074	0.006961	0.000853	0.000093
		α	10^{-1}	10^{-2}	10^{-3}	10^{-4}
		With noise δ	10^{-3}	10^{-6}	10^{-9}	10^{-12}
No reg.	Midpoint rect.	PWC	0.139125	0.014874	0.001583	0.000165
		PWL	0.025508	$1.03 \cdot 10^{-3}$	$2.5 \cdot 10^{-5}$	$2.2 \cdot 10^{-6}$
	Gauss	PWC	0.139899	0.014874	0.001616	0.000165
		PWL	0.011047	$5.5 \cdot 10^{-4}$	$2.3 \cdot 10^{-5}$	$3.4 \cdot 10^{-7}$
With regularisation	Midpoint rect.	PWC	0.137952	0.014855	0.001568	0.000162
		α	10^{-2}	10^{-4}	10^{-6}	10^{-8}
		PWL	$4.0 \cdot 10^{-3}$	$1.63 \cdot 10^{-4}$	$9.7 \cdot 10^{-6}$	$3.6 \cdot 10^{-7}$
		α	10^{-4}	10^{-4}	10^{-5}	10^{-6}
	Gauss	PWC	0.138185	0.014863	0.001551	0.000162
		α	10^{-2}	10^{-4}	10^{-6}	10^{-8}
		PWL	0.005295	$2.1 \cdot 10^{-4}$	$7.5 \cdot 10^{-6}$	$1.8 \cdot 10^{-7}$
		α	10^{-4}	10^{-4}	10^{-5}	10^{-6}

(*Continued*)

Table 3.1.   (*Continued*)

Notes: The above calculations demonstrate that the use of Lavrentiev's α-regularisation makes it possible to significantly reduce the level of error in the numerical solution of irregular integral equations of the first kind for noisy data, especially when, for a given noise level, the self-regularisation property characteristic of Volterra integral equations of the first kind has not yet manifested itself. At a high noise level, calculations without α-regularisation became unstable or produced a large error.

For PWL approximation, using both Gaussian quadratures and average rectangles, a notable improvement in the result is observed when matching the grid step of nodes and the level of noise added to the kernel and the right-hand side. This is due to the well-known property of self-regularisation when matching the grid step of nodes with the noise level. To find the optimal step and regularisation parameter α with a known noise level of the initial data of the equation, you can use standard methods of minimising a convex function (for example, the Fibonacci method).

As a footnote, it is worth mentioning that the Laurentiev regularisation is essential for addressing a wide range of practical problems, such as image reconstruction, tomography, signal processing, and inverse scattering problems. It provides a systematic framework for handling ill-posed integral equations and operators, making it an important and valuable tool for researchers and practitioners in the field.

Calculations were performed using the quadrature formulas of average rectangles and Gaussian quadratures (over three nodes).

3.3   On the Role of *A Priori* Estimates in the Nonlocal Continuation of Solutions Method with Respect to a Parameter

This section focuses on an iterative method, also known as the homotopy analysis method (HAM), for continuing solutions with respect to a parameter. The research specifically examines the nonlocal scenario where the parameter is within a segment of the real axis. Essentially, an iterative approach is developed for continuing the solution for a linear equation in Banach spaces with a linear operator that depends on the parameter and satisfies the Lipschitz condition concerning the parameter. The findings are then extended to nonlinear equations in Banach spaces, with the assumption that the nonlinear mapping is dependent on a real parameter. An iterative approach is established for continuing the solution with respect to the parameter using the Newton–Kantorovich method. The study leverages the availability of *a priori* estimates of solutions, allowing for the construction of a solution for any parameter value.

The parameter continuation method has been developed since the 20th century by many outstanding mathematicians, including S. Bernstein, G. Weil, G. Levi, J. A. Poincaré, A. M. Lyapunov, and A. I. Nekrasov. The method was used to prove constructive existence theorems, made it possible to solve a number of complex applied problems, and stimulated the development of new areas of nonlinear analysis. The stages of these studies are reflected in the review by Lyusternik (1956). Recently, the parameter continuation method has received further development in accordance with the latest achievements of computer implementations, see, for example, the works of He (1999), Fedorov *et al.* (2019), and Noeiaghdam *et al.* (2020a), and their references.

The method of nonlocal continuation of solutions with respect to a parameter and its various versions are important because such an approach can provide accurate and reliable solutions to problems that are difficult or impossible to solve using traditional analytical or numerical methods. They can also be used to investigate the behaviour of nonlinear systems and gain insights into the underlying dynamics of these systems.

Some variants of this method are called the method of homotopic perturbations. An increasing efficiency of the method has been observed both in nonlinear analysis and in a number of applied areas of modern science and technology. In this connection, its rigorous justification with an indication of convergence bounds with respect to the parameters is an essential task we focus on. The practical advantage of the method is based on the simplicity of its algorithmisation. It should be especially noted that the convergence of the corresponding methods of successive approximations of parametric branches of solutions, as a rule, was established only in a local situation. In the theory of global reversibility of operators with parameters, there are only some partial results, see, for example, Sidorov *et al.* (2020b) and Trenogin (1996). The basics of analytical, topological, variational, group, and approximate methods in the theory of continuation of solutions for nonlinear equations in irregular cases are given in a number of papers (see, for example, Sidorov and Trufanov, 2009; Sidorov, 1984; Sidorov *et al.*, 2010). A detailed bibliography can be found in the monographs by Korpusov and Panin (2016) and Vainberg and Trenogin (1964). The approach is versatile and can be applied to a variety of nonlinear problems, including those with multiple variables, boundary conditions, and initial conditions. This makes it a valuable tool for researchers and engineers working on a wide range of problems in various fields.

In this section, we consider the method of successive approximations of parametric solutions when the parameter belongs to a segment of the real axis. In the proof of convergence, the existence of *a priori* estimates of the solution is used. Note that *a priori* estimates are usually used in works on global solvability by Korpusov and Panin (2016).

The section is structured as follows. Section 3.3.1 presents the method of nonlocal continuation of solution for a linear equation. Next, the results of Section 3.3.1 are generalised and applied to nonlinear equations using the Newton–Kantorovich method. Section 3.3.2 considers the application of the method of continuation by parameter in the choice of the initial approximation of solution for a linear equation. The prospects of the parameter continuation method in the irregular case are discussed in the concluding section.

3.3.1   *On the continuation by parameter of solutions of linear operator equations*

Consider the linear equation

$$B(\lambda)x = f. \tag{3.35}$$

Here, $\lambda \in [0, \rho] \subset \mathbb{R}^1$, $B : [0, \rho] \to \mathcal{L}(X \to Y)$ is a linear bounded operator, X, Y are Banach spaces, and $f \in Y$. The goal is to construct the solution $x(\lambda)$ at the points $\lambda_i = ih$, $i = 1, \ldots, N$, where $h = \frac{\rho}{N}$, and specify a way to compute the inverse operators $B^{-1}(\lambda_i)$.

Theorem 3.3.1. *Let there exist a bounded inverse operator $B^{-1}(0)$ and the Lipschitz condition $\|B(\lambda) - B(\mu)\| \leq l|\lambda - \mu|$ be satisfied at λ, μ of the segment $[0, \rho]$. Let*

$$\|B(\lambda)x\|_Y \geq \gamma(\lambda)\|x\|\|_X \tag{3.36}$$

when $\lambda \in [0, \rho]$ and $x \in X$. Then, there exists $N_0 < \infty$ such that, at the points $\lambda_i = ih$, $i = 1, \ldots, N$, where $h = \frac{\rho}{N}$ and $N \geq N_0$, there exist bounded inverse operators $B^{-1}(\lambda_i)$, which can be constructed using successive approximations:

$$B^{-1}(h) = \lim_{n \to \infty} \sum_{k=0}^{n} (B^{-1}(0)(B(0) - B(h))^k)B^{-1}(0), \tag{3.37}$$

$$B^{-1}(jh) = \lim_{n \to \infty} \sum_{k=0}^{n} \left(B^{-1}((j-1)h)(B((j-1)h) - B(jh)) \right)^k$$

$$\times B^{-1}((j-1)h), \ j = 2, \dots, N. \tag{3.38}$$

Proof. Since $B^{-1}(0) \in \mathcal{L}(Y \to X)$ and the *a priori* estimate (3.36) is fulfilled,

$$\|B^{-1}(0)\| \leq 1/\gamma.$$

Fix $q < 1$. Using the identity

$$B(h) = B(0)[I + B^{-1}(0)(B(h) - B(0))],$$

where $h \leq \frac{\gamma}{l}q$, let us construct linear inverse operator (3.37) based on the known properties of linear operators.

To prove the convergence of sequence (3.37) in the sense of the norm of the Banach space $\mathcal{L}(Y \to X)$, it is not difficult to establish, with the help of the Lipschitz inequality, the estimate $\|B^{-1}(0)(B(0) - B(h))\| \leq q < 1$ at $h \leq \frac{\gamma}{l}q$. Note that in this case, by virtue of the *a priori* estimate (3.36), we obviously obtain the estimate

$$\|B^{-1}(h)\| \leq 1/\gamma.$$

The proof is completed by applying the method of mathematical induction. Indeed, let $\||B(ih)\||^{-1} \leq 1/\gamma$, $i = 1, \dots, j - 1$. Then, $B(jh) = B((j-1)h)[I - (B^{-1}((j-1)h)(B((j-1)h) - B(jh))]$, where $\||B^{-1}((j-1)h)(B((j-1)h) - B(jh))\|| \leq \frac{1}{\gamma}lh \leq q < 1$. Hence, $B^{-1}(jh) \in \mathcal{L}(Y \to X)$, and by virtue of (3.36),

$$\||B^{-1}(jh)\|| \leq \frac{1}{\gamma}.$$

Hence, the sought inverse operators exist and are constructed using formula (3.38), where the convergence to a linear bounded operator follows from the completeness of the space $\mathcal{L}(Y \to X)$. The theorem is proved. □

Corollary 3.3.2. *Let the conditions of Theorem 3.3.1 be satisfied. Then, for sufficiently small $h > 0$ in the solutions of equation (3.35) at the points $\{ih\}_{i=1}^{N}$ of $[0, \rho]$, the recurrence formulae are*

$$x(0) = B^{-1}(0)f,$$

$$x(h) = \lim_{n \to \infty} \sum_{k=0}^{n} (I + B^{-1}(0)(B(h) - B(0)))^k x(0),$$

$$x(jh) = \lim_{n\to\infty} \sum_{k=0}^{n} \left(I + B^{-1}((j-1)h)(B(jh) - B((j-1)h))\right)^k x((j-1)h),$$

$$(3.39)$$

$$j = 2, 3, \ldots, N.$$

Formula (3.38) can be employed to compute inverse operators.

Remark 3.3.3. Since, by virtue of Theorem 3.3.1, the operators $B(jh)$ are continuously reversible, to construct a sequence of elements $x(0), x(h)$, $x(2h), \ldots,$ in addition to successive iterations of (3.39), other known methods can be used.

Corollary 3.3.4. *Let the conditions of Theorem 3.3.1 be satisfied. Then, the operator $B(\lambda)$ has a bounded inverse at $\forall \lambda \in [0, \rho]$.*

Proof. The proof is obvious since if the conditions of Theorem 3.3.1 are satisfied, the set of points λ_i, where the operator $B(\lambda)$ has a bounded inverse, is open and closed on the segment $[0, \rho]$. □

Newton–Kantorovich method of convergence. Let us focus on the convergence of successive approximations in the Newton–Kantorovich method in the case of continuation of solution by a parameter.

Let us consider the following equation:

$$F(x, \lambda) = f,$$

$$(3.40)$$

where $F : X \times [0, \rho] \to Y$ is a nonlinear mapping, X, Y are Banach spaces, $\lambda \in [0, \rho]$, and $f \in Y$. Let us apply the abstract formulation of Picard's method of successive approximations.

Let the following conditions be satisfied:

(1) The mapping $F(x, \lambda)$ and its derivatives $F_x(x, \lambda)$, $F_\lambda(x, \lambda)$ are defined at $x \in X$, and $\lambda \in [0, \rho]$, are continuous, and satisfy the Lipschitz condition on x and λ in the region $D = \{\lambda, x | \lambda \in [0, \rho], ||x - x_0|| < R\}$, where the element x_0 satisfies the equation

$$F(x, 0) = f.$$

$$(3.41)$$

(2) There exists a bounded inverse operator

$$F_x^{-1}(x_0, 0) \in \mathcal{L}(Y \to X).$$

The solution $x(\lambda)$ of equation (3.35) should be constructed at the points $\lambda_i = ih$, $i = 1, 2, \ldots, N$ at $h = \frac{\rho}{N}$. If h is small enough, then $\forall \lambda \in [0, h]$ based on the theorem of implicit mapping (Sidorov and Sidorov, 2017). The sequence $\{x_n(\lambda)\}$, computed uniquely from the linear equations

$$F_x(x_0, 0)[x_n(\lambda) - x_{n-1}(\lambda)] + F(x_{n-1}(\lambda), \lambda) = f \qquad (3.42)$$

at the initial approximation $x_0(\lambda) = x_0$, converges to the solution $x(\lambda)$ of equation (3.40). Moreover, the function $x(\lambda)$ will satisfy the Lipschitz condition by virtue of condition 1. Thus, $x(h) = \lim_{n \to \infty} x_n(h)$, where $\{x_n(h)\}$ is constructed using formula (3.42) of the Newton–Kantorovich method. In order to guarantee the convergence of this method for further continuation of the solution by the parameter λ, let us introduce a further condition:

(3) Let the estimate $|||x - x_0||| \leq R$ be an *a priori* estimate of the desired continuous solution of equation (3.40) at $\lambda \in [0, \rho]$ and $||F_x(x, \lambda)u|| \geq \gamma ||u|||$ when $||x - x_0|| \leq R$ and $\lambda \in [0, \rho]$.

Then, $||F_x^{-1}(x(h), h)|| \leq \frac{1}{\gamma}$, and we can construct the element $x(2h)$ by solving the sequence of equations

$$F_x(x(h), h)[x_n(2h) - x_{n-1}(2h)] + F(x_{n-1}(2h), 2h) = f, \quad n = 1, 2, \ldots$$

at the initial approximation $x_0(2h) = x(h)$. The elements $x_n(2h)$ remain in the region of attraction of the method, and $\lim_{n \to \infty} x_n(2h) = x(2h)$ based on Kantorovich's theorem.

Thus, condition 3 allows us to ensure that the solution of equation (3.40) can be continued by the parameter λ. Using the Newton–Kantorovich method, we can sequentially compute the elements $x(ih)$, $i = 1, \ldots, N$, where $h = \frac{\rho}{N}$ and is sufficiently small.

3.3.2 Application of the parameter continuation method in selecting the initial approximation of the solution of a linear equation

Let us consider the linear equation

$$Ax = f, \qquad (3.43)$$

where $A \in \mathcal{L}(X \to Y)$, $f \in Y$.

Let's introduce an auxiliary linear continuous inverse operator, $B \in \mathcal{L}(X \to Y)$. The equation

$$Bx = f \qquad (3.44)$$

has a single solution, $x_0 = B^{-1}f$, which we assume to be constructed. We note the possibility of a wide choice of the operator B since $||B - A||$ can be as large as desired.

Let us introduce the linear equation

$$B(\lambda)x = f, \qquad (3.45)$$

where $B(\lambda) = B + \lambda(A - B)$, $\lambda \in [0, 1]$, and

$$||B(\lambda_1) - B(\lambda_2)|| = ||B - A|| |\lambda_1 - \lambda_2| \qquad (3.46)$$

for all λ_1, λ_2.

The operator $B(\lambda)$ is called a homotopy for the operators B and A (see He, 1999, 2003). We assume that the operator B is chosen in such a way that an *a priori* estimate can be obtained as

$$||B(\lambda)x||_Y \geq \gamma ||x||_X \qquad (3.47)$$

for all $\lambda \in [0, 1]$, $\gamma > 0$. Then, obviously

$$||B^{-1}||_{\mathcal{L}(Y \to X)} \leq \frac{1}{\gamma},$$

and equation (3.45), for $0 < \lambda \leq 1$, can have at most one solution.

Fix $0 < q < 1$ and choose $h \leq q\frac{\gamma}{||B-A||}$. Then, at $0 < \lambda \leq h$, the sequence $\{x_n(\lambda)\}_{n=1}^{\infty}$, defined uniquely based on the linear equation

$$Bx_n(\lambda) + \lambda(B - A)x_{n-1}(\lambda) = f \qquad (3.48)$$

at the initial approximation $x_0(\lambda) \equiv x_0$, converges uniformly to the solution $x(\lambda)$ of equation (3.45). We can continue this process further and construct a solution to the original equation (3.43) at λ on the segment $[0, 1]$.

Indeed, let us choose on the segment $[0, 1]$ the points $\lambda_i = ih_0$, $i = 0, 1, \ldots, N$, $h_0 = \frac{1}{N} \leq h$. Then, by Theorem 3.3.1, all operators $B(\lambda_i)$ will be continuously reversible, and when solving the ith equation $B(\lambda_i)x_i = f$, $i = 2, 3, \ldots, N$, we can use as an initial approximation solution, x_{i-1}, of the previous equation. Thus, to find x_1, the solution x_0 of the auxiliary equation (3.44) is used. We have used the above method to solve the Volterra integral equations (Noeiaghdam *et al.*, 2020a).

The method of homotopy perturbations has also been successfully used in the numerical solution of some classes of nonlinear differential equations (He, 1999; Fedorov *et al.*, 2019).

In this section, we presented and discussed the method of continuation by a parameter in the regular case when the operator $B(0)$ (respectively, the operator $F_x(x_0, 0)$) is continuously reversible. If the operator $B(0)$ is not reversible and the estimate $||B(\lambda)x|| \geq c\lambda^n||x||$ at $\lambda \in (0, \rho]$, then the solution of the considered nonlinear equation can have at the point $\lambda = 0$ a pole of nth order. If the operator $F_x(x_0, 0)$ is not reversible, then the point $\lambda = 0$ will be the branching point of the solution, and the equation can have several solutions $x(\lambda)$ satisfying the condition $\lim_{\lambda \to 0} x(\lambda) = x_0$. The question of choosing the optimal step of continuation on the parameter requires a separate consideration that takes the specificity of the problem into account.

The development of approximate methods for constructing solutions in irregular cases is of special interest and requires the involvement of subtle and sophisticated methods, such as the method of skeleton decomposition of operators, group symmetry, bifurcation theory, and contemporary regularisation methods of ill-posed problems.

3.4 The Adomian Decomposition for Numerical Solution of the Volterra Equation with Discontinuous Kernels Using the CESTAC Method

This section presents an overview of using stochastic arithmetic (SA) to validate the numerical results of linear and nonlinear Volterra integral equations with discontinuous kernels. The traditional absolute error is replaced with SA and new conditions to improve method efficiency. The CESTAC (from french; Contrôle et Estimation STochastique des Arrondis de Calculs) method and CADNA (Control of Accuracy and Debugging for Numerical Applications) library are utilised, offering advantages over floating-point arithmetic (FPA). The convergence analysis of the Adomian decomposition method (ADM) is proven, and the CESTAC method's main theorem is presented, allowing for a new termination criterion. Examples in linear and nonlinear cases are compared to demonstrate the accuracy of the new method. The results presented in this section were obtained jointly with S. Noeiaghdam and V. Sizikov.

Volterra integral equations of the first kind with discontinuous kernels belong to the class of *ill-posed* problems. The discontinuous kernels introduce fundamental difficulties in the theory of nonlinear Volterra equations of the first kind: there is a loss of uniqueness of solutions, solutions may blow up, or branching phenomena may occur. The existence of a continuous solution depending on free parameters and sufficient conditions for the existence of a unique continuous solution for systems of Volterra integral equations of the first kind with discontinuous kernels were derived by Sidorov (2013b). For a comprehensive review of the results in this field, readers may refer to the book by Sidorov (2014b).

The ADM is an iterative and applicable method for solving various problems such as the Klein–Gordon equation (Saelao and Yokchoo, 2020), the Trini–Biswas equation (González-Gaxiola, 2019), problems of boundary layer convective heat transfer (Daoud and Khidir, 2018), integral equations (IEs) of the first and second kinds with hypersingular kernels (Novin *et al.*, 2018; Mahmoudi, 2014), the Volterra integral form of Lane–Emden equations with initial values and boundary conditions (Wazwaz *et al.*, 2013), Cauchy IEs of the first kind (Bougoffa *et al.*, 2013), linear and nonlinear IEs (Wazwaz, 2011), and partial differential equations (Wazwaz, 2009).

These methods, along with many other methods, of solving the Volterra integral equations are based on FPA. Thus, in order to demonstrate the accuracy of the numerical results, the authors apply the absolute error as follows:

$$|y(t) - y_n(t)| < \varepsilon, \tag{3.49}$$

where $y(t)$ and $y_n(t)$ are the exact and approximate solutions, respectively. However, there can be certain disadvantages. Condition (3.49) depends on the value ε as well as the exact solution. But we do not know the optimal value of ε, and in many cases, we do not have the exact solution to compare the results. If we choose small values for ε, we will require extra iterations, whereas if we set large values, then the numerical process will stop too soon, and we will not be able to obtain accurate results.

Thus, in order to demonstrate the efficiency of the numerical procedures, instead of condition (3.49), we apply the following termination criterion:

$$|y_n(t) - y_{n+1}(t)| = @.0, \tag{3.50}$$

which depends on two successive approximations, $y_n(t)$ and $y_{n+1}(t)$. On the right-hand side, we have the informational zero @.0, which shows that the number of common significant digits (NCSDs) between two successive approximations is zero.

Because of these problems, we introduce SA as an alternative to FPA (Chesneaux, 1992). In SA, we apply the CESTAC method, and instead of the absolute error, we use the termination criterion based on two successive approximations. So, we do not require the exact solution. Also, the informational zero @.0 is used instead of ε. The numerical algorithm will be stopped when the NCSDs of two successive approximations equal zero. Furthermore, the CESTAC method can be implemented on the CADNA library using the Linux operating system, in which its codes must be written in ADA, FORTRAN, or C/C++ codes (Chesneaux, 1990). Using the CESTAC method and the CADNA library, we can find the optimal approximation, error, and iteration of the numerical procedure (Jézéquel and Mecanique, 2006; Chesneaux and Jézéquel, 1998). The CESTAC method was studied by Laporet and Vignes for the first time, following which researchers from LIP6, the computer science laboratory at Sorbonne University in Paris, France,[1] extended this method by developing the CADNA library (Vignes, 2004, 1993; Graillat *et al.*, 2011).

This study applies the ADM for solving linear and nonlinear Volterra integral equations with discontinuous kernels and validates the numerical results using the CESTAC method and the CADNA library. Therefore, we will be able to find the optimal approximation, the optimal error, and the optimal iteration of the ADM for solving equation (3.52). The uniqueness theorem, the error theorem, and the convergence theorem of the ADM are proved. Additionally, the main theorem of the CESTAC method is discussed. Based on this theorem, we can apply the new termination criterion instead of the absolute error. Several examples are solved, and the CESTAC method is applied to validate the results and find the optimal results of the ADM for solving the mentioned problem.

3.4.1 *Stochastic arithmetic and the CESTAC method*

The CESTAC method is based on a probabilistic approach of round-off error propagation, which can help us replace FPA with random arithmetic. The parallel implementation is one of the favorable aspects of this method. Applying this method, k runs of the computer program can be done in parallel. Thus, a new arithmetic, which we call SA, is defined. For definitions and properties of SA, readers may refer to Chesneaux (1992). In order to apply the CESTAC method, we should substitute SA for FPA. Thus, we will be able to run each arithmetical operation k times synchronously before running the next operation. The entire process should be implemented using the CADNA library. During the run, the CADNA library can find the

---

[1] https://www-pequan.lip6.fr/.

NCSDs of each result, and if the result is zero, then the CADNA library will be stopped by showing the informational zero @.0. Thus, each result can appear as a random variable.

If we produce the representable values on a computer and collect them in B, then $S^* \in B$ can be written for $s^* \in \mathbb{R}$, with α mantissa bits of the binary FPA, as

$$S^* = s^* - \rho 2^{E-\alpha}\phi, \tag{3.51}$$

where $\rho, 2^{-\alpha}\phi$, and E are the sign, the missing segment of the mantissa, and the binary exponent of the result, respectively. Also, we know that for $\alpha = 24, 53$, the numerical results can be produced in single and double precision (Chesneaux and Jézéquel, 1998; Chesneaux, 1992; Jézéquel and Mecanique, 2006). By assuming ϕ to be a casual variable that is uniformly distributed on $[-1, 1]$, we will be able to introduce perturbations in the last mantissa bit of s^*. Then, the mean (μ) and the standard deviation (σ) values can be produced for the results of S^*, which play an important role in determining the precision of S^*. If we repeat the process for k times, we will have the quasi-Gaussian distribution on $S_i^*, i = 1, \ldots, k$ and we will have equality between μ and the exact s^*.

Algorithm 1 shows the process step by step, where τ_δ is the value of the T distribution when the confidence interval is $1 - \delta$, with $k - 1$ degrees of freedom (Graillat *et al.*, 2011; Vignes, 1993; Chesneaux and Jézéquel, 1998).

Algorithm 1:

Step 1. Produce k samples of S^* in the form of $\Phi = \{S_1^*, S_2^*, \ldots, S_k^*\}$ by making perturbation on the last bit of mantissa.

Step 2. Calculate $\tilde{S}^* = \frac{\sum_{i=1}^k S_i^*}{k}$.

Step 3. Find $\sigma^2 = \frac{\sum_{i=1}^k (S_i^* - \tilde{S}^*)^2}{k-1}$.

Step 4. Apply $C_{\tilde{S}^*, S^*} = \log_{10} \frac{\sqrt{k}|\tilde{S}^*|}{\tau_\delta \sigma}$ to find the NCSDs between S^* and $\tilde{S}^*$.

Step 5. Show $S^* = $ @.0 if $\tilde{S}^* = 0$, or $C_{\tilde{S}^*, S^*} \leq 0$.

In order to apply the CESTAC method, we do not need to apply the mentioned algorithm directly with commonly used software, such as MATLAB, Mathematica, and Maple. Instead, this method can be implemented using the CADNA library, for which we need to write the CADNA codes using C, C++, FORTRAN, or ADA codes (Chesneaux,

1990). Then, the CESTAC method can be automatically implemented in the numerical procedures.

The application of the CESTAC method and the CADNA library has the following advantages over mathematical methods based on FPA:

- In FPA, the numerical algorithm may stop too soon before producing accurate results.
- Generally, FPA depends on the absolute error, for which we need the exact solution, whereas in the CESTAC method, we do not need the exact solution.
- In some cases, the absolute error depends on the positive small value ε whose optimal value is not known to be used. In the CESTAC method, this value is not required.
- In the CESTAC method, the algorithm will stop with the optimal iteration, whereas in FPA, extra iterations might be produced without improving the accuracy of the results.
- In the CESTAC method, we can identify the optimal values, such as optimal iteration, approximation, and error, which is not possible using FPA.

The following are some sample codes of the CADNA library:

```
# include <cadna.h>
cadna_init(-1);
main()
{
double_st Parameter;
do
{
    Write the main program here;
printf(" %s ",strp(Parameter));
}
while(u[n]-u[n-1]!=0);
cadna_end();
}
```

Consider the following nonlinear Volterra integral equation of the second kind with a discontinuous kernel with respect to continuous $y(t)$:

$$y(t) = x(t) + \sum_{j=1}^{m'} \int_{\beta_{j-1}(t)}^{\beta_j(t)} k_j(t,\tau)F(y(\tau))d\tau, \qquad (3.52)$$

where

$$0 =: \beta_0(t) < \beta_1(t) < \cdots < \beta_{m'-1}(t) < \beta_{m'}(t) := t, \quad for \ t \in (0,T),$$

and $\beta_j(0) = 0$. Also, $\forall t \in J = [0,T]$, we assume that $x(t)$ is bounded and $k_j(t,\tau)$ is discontinuous along continuous curves $\beta_j(t)$, $j = 0,1,\ldots,m'$ such that $|k_j(t,\tau)| < M_j$, $\forall 0 \leq \tau \leq t \leq T$ and the nonlinear term $F(y)$ satisfies in the Lipschitz continuous such that $|F(y) - F(z)| \leq L|y - z|$.

The ADM assumes that the unknown function $y(t)$ can be constructed as an infinite series of the form

$$y(t) = \sum_{i=0}^{\infty} y_i(t), \tag{3.53}$$

and the Adomian polynomials (El-kalla, 2005) can be obtained in the following form:

$$A_n = F(P_n) - \sum_{j=0}^{n-1} A_j, \tag{3.54}$$

where $P_n = \sum_{i=0}^{n} y_i(t)$ shows the partial sum. Then, we have

$$y_0(t) = x(t),$$

$$y_i(t) = \sum_{j=1}^{m'} \int_{\beta_{j-1}(t)}^{\beta_j(t)} k_j(t,\tau) A_{i-1} d\tau, \quad i \geq 1. \tag{3.55}$$

Also, the nonlinear term $F(y)$ can be decomposed into an infinite series of polynomials given by

$$F(y) = \sum_{n=0}^{\infty} A_n, \tag{3.56}$$

where the Adomian polynomials are as follows:

$$A_n = \left(\frac{1}{n!}\right)\left(\frac{d^n}{d\lambda^n}\right)\left[F\left(\sum_{i=0}^{\infty} \lambda^i y_i\right)\right]_{\lambda=0}. \tag{3.57}$$

The following theorems show the uniqueness, convergence, and error of the method. The well-known contraction mapping principle is applied to prove them.

Lemma 3.4.1. *If we apply the ADM for solving equation (3.52), the obtained solution will be unique whenever* $0 < \eta < 1$, *where* $\eta = L\sum_{j=1}^{m'} M_j(\beta_j - \beta_{j-1})$.

Proof. See El-kalla (2005). □

Theorem 3.4.2. *The series solution* (3.53) *for solving equation* (3.52) *using the ADM converges if* $0 < \eta < 1$ *and* $|y_1| < \infty$.

Proof. Let $(C[J], \|.\|)$ be the Banach space of all continuous functions on J such that $\|f(t)\| = \max_{t \in J} |f(t)|$. Let $\{P_n\}$ be a sequence of partial sums, where P_n and P_m are arbitrary partial sums with $n \geq m$. We must prove that $\{P_n\}$ is a Cauchy sequence in the Banach space:

$$\|P_n - P_m\| = \max_{\forall \in J} |P_n - P_m| = \max_{\forall \in J} \left| \sum_{i=m+1}^{n} y_i(t) \right|$$

$$= \max_{\forall \in J} \left| \sum_{i=m+1}^{n} \sum_{j=1}^{m'} \int_{\beta_{j-1}(t)}^{\beta_j(t)} k_j(t,\tau) A_{i-1} d\tau \right|$$

$$= \max_{\forall \in J} \left| \sum_{j=1}^{m'} \int_{\beta_{j-1}(t)}^{\beta_j(t)} k_j(t,\tau) \sum_{i=m}^{n-1} A_i d\tau \right|.$$

Using equation (3.54), we can write $\sum_{i=m}^{n-1} A_i = F(P_{n-1}) - F(P_{m-1})$. Then,

$$\|P_n - P_m\| = \max_{\forall \in J} \left| \sum_{j=1}^{m'} \int_{\beta_{j-1}(t)}^{\beta_j(t)} k_j(t,\tau) [F(P_{n-1}) - F(P_{m-1})] d\tau \right|$$

$$\leq \max_{\forall \in J} \sum_{j=1}^{m'} \int_{\beta_{j-1}(t)}^{\beta_j(t)} |k_j(t,\tau)| |F(P_{n-1}) - F(P_{m-1})| d\tau$$

$$= \eta \|P_{n-1} - P_{m-1}\|,$$

and for $n = m + 1$, we have

$$\|P_{m+1} - P_m\| \leq \eta \|P_m - P_{m-1}\| \leq \eta^2 \|P_{m-1} - P_{m-2}\| \leq \cdots \leq \eta^m \|P_1 - P_0\|.$$

Applying the triangle inequality, we have

$$\|P_n - P_m\| \leq \|P_{m+1} - P_m\| + \|P_{m+2} - P_{m+1}\| + \cdots + \|P_n - P_{n-1}\|$$

$$\leq [\eta^m + \eta^{m+1} + \cdots + \eta^{n-1}] |P_1 - P_0\|$$

$$\leq \eta^m \left(\frac{1 - \eta^{n-m}}{1 - \eta} \right) \|y_1(t)\|.$$

Since $0 < \eta < 1$, we can write $1 - \eta^{n-m} < 1$, and we have

$$\|P_n - P_m\| \leq \frac{\eta^m}{1 - \eta} \max_{\forall t \in J} |y_1(t)|. \tag{3.58}$$

We know that $x(t)$ is bounded and $|y_1| < \infty$. So, $\|P_n - P_m\|$ converges to zero as m approaches infinity. This shows that P_n is a Cauchy sequence in $C[J]$, and the series converges. $\qquad\square$

Theorem 3.4.3. *If we apply series solution* (3.53) *to solve equation* (3.52), *the maximum absolute error truncation can be obtained as follows:*

$$\max \left| y(t) - \sum_{i=0}^{m} y_i(t) \right| \leq \frac{k\eta^{m+1}}{L(1 - \eta)},$$

where $k = \max_{\forall t \in J} |F(x(t))|$.

Proof. Applying inequality (3.58) and Theorem 3.4.2 leads to

$$\|P_n - P_m\| \leq \frac{\eta^m}{1 - \eta} \max_{\forall t \in J} |y_1(t)|.$$

If n approaches ∞, then s_n will approach $y(t)$ and

$$|y_1(t)| \leq \sum_{j=1}^{m'} M_j(\beta_j - \beta_{j-1}) \max |F(y_0)|,$$

and

$$\|y(t) - P_m\| \leq \frac{\eta^{m+1}}{L(1 - \eta)} \max_{\forall t \in J} |F(x(t))|.$$

Finally, the maximum error in J can be obtained as

$$\max_{\forall t \in J} |y(t) - \sum_{i=0}^{m} y_i(t)| \leq \frac{k\eta^{m+1}}{L(1 - \eta)}. \qquad\square$$

Introducing an auxiliary parameter and differentiating with respect to it for calculating the initial approximations have been effectively employed in other nonlinear problems. Here, readers may refer to Sidorov (2001).

Definition 3.4.1 (Chesneaux and Jézéquel (1998) and Chesneaux (1992)). The NCSDs for two real numbers r_1, r_2 can be obtained as follows:

(1) for $r_1 \neq r_2$,

$$C_{r_1,r_2} = \log_{10} \left| \frac{r_1 + r_2}{2(r_1 - r_2)} \right| = \log_{10} \left| \frac{r_1}{r_1 - r_2} - \frac{1}{2} \right|; \qquad (3.59)$$

(2) for all real numbers r_1, $C_{r_1,r_1} = +\infty$.

Theorem 3.4.4. *Let $y(t)$ and $y_n(t)$ be the exact and numerical solutions of problem (3.52), respectively, where $y_n(t)$ is obtained by using the ADM. We have*

$$C_{y_n(t),y_{n+1}(t)} \simeq C_{y_n(t),y(t)}, \qquad (3.60)$$

where $C_{y_n(t),y(t)}$ denote the NCSDs of $y_n(t), w(t)$ and $C_{y_n(t),y_{n+1}(t)}$ are the NCSDs of two successive iterations $y_n(t), y_{n+1}(t)$.

Proof. Using Definition 3.4.1, we get

$$C_{y_n(t),y_{n+1}(t)}$$

$$= \log_{10} \left| \frac{y_n(t)}{y_n(t) - y_{n+1}(t)} - \frac{1}{2} \right|$$

$$= \log_{10} \left| \frac{y_n(t)}{y_n(t) - y_{n+1}(t)} \right| + \log_{10} \left| 1 - \frac{1}{2y_n(t)} (y_n(t) - y_{n+1}(t)) \right|$$

$$= \log_{10} \left| \frac{y_n(t)}{y_n(t) - y_{n+1}(t)} \right| + \mathcal{O}\big(y_n(t) - y_{n+1}(t)\big)$$

$$= \log_{10} \left| \frac{y_n(t)}{(y_n(t) - y(t)) - (y_{n+1}(t) - y(t))} \right|$$

$$\quad + \mathcal{O}\Big[(y_n(t) - y(t)) - (y_{n+1}(t) - y(t))\Big]$$

$$= \log_{10} \left| \frac{y_n(t)}{(y_n(t) - y(t)) \left[1 - \frac{y_{n+1}(t) - y(t)}{y_n(t) - y(t)}\right]} \right| + \mathcal{O}(E_n) + \mathcal{O}(E_{n+1})$$

$$= \log_{10} \left| \frac{y_n(t)}{y_n(t) - y(t)} \right| - \log_{10} \left| 1 - \frac{y_{n+1}(t) - y(t)}{y_n(t) - y(t)} \right| + \mathcal{O}\left(\frac{\eta^{n+1}}{1 - \eta}\right)$$

$$= \log_{10} \left| \frac{y_n(t)}{y_n(t) - y(t)} \right| - \log_{10} \left| 1 - \frac{y_{n+1}(t) - y(t)}{y_n(t) - y(t)} \right| + \mathcal{O}\left(\frac{\eta^{n+1}}{1 - \eta}\right).$$

$$(3.61)$$

Also,

$$C_{y_n(t),y(t)} = \log_{10} \left| \frac{y_n(t)}{y_n(t) - y(t)} - \frac{1}{2} \right|$$

$$= \log_{10} \left| \frac{y_n(t)}{y_n(t) - y(t)} \right| + \mathcal{O}(y_n(t) - y(t))$$

$$= \log_{10} \left| \frac{y_n(t)}{y_n(t) - y(t)} \right| + \mathcal{O}\left(\frac{\eta^{n+1}}{1 - \eta} \right). \tag{3.62}$$

Applying equations (3.61) and (3.62), we have

$$C_{y_n(t),y_{n+1}(t)} = C_{y_n(t),y(t)} - \log_{10} \left| 1 - \frac{y_{n+1}(t) - y(t)}{y_n(t) - y(t)} \right| + \mathcal{O}\left(\frac{\eta^{n+1}}{1 - \eta} \right).$$

From Theorem 3.4.3, we can write $\frac{y_{n+1}(t)-y(t)}{y_n(t)-y(t)} = \frac{\mathcal{O}\left(\frac{\eta^{n+2}}{1-\eta} \right)}{\mathcal{O}\left(\frac{\eta^{n+1}}{1-\eta} \right)} = \mathcal{O}(\eta)$. Thus, for n large enough, we get

$$C_{y_n(t),y_{n+1}(t)} \simeq C_{y_n(t),y(t)}.$$

$\square$

It is worth highlighting that when n increases, the NCSDs between two sequential results obtained from the algorithm are almost equal to the NCSDs of the nth iteration and the exact solution at the given point t, which means that for an optimal index such as $n = n_opt$, when $y_n(t) - y_{n+1}(t) = @.0$, $y_n(t) - y(t) = @.0$.

3.4.2   *Numerical results*

In this section, we apply the ADM for solving the mentioned examples. The numerical results are obtained using both FPA and SA. In FPA, the numerical algorithm depends on the value ε. Also, the numbers of iterations for different values of ε are obtained. It is obvious that for small values of ε, the algorithm cannot be stopped, and we will have many iterations without improving the accuracy of the results. Conversely, for large values of ε, the algorithm will be stopped too soon without providing accurate results. In SA, along with the application of the CESTAC method and the CADNA library, we can find the optimal results and the optimal iteration and error of the ADM for solving the Volterra integral equations in linear and nonlinear forms with discontinuous kernels. Thus, we can clearly see the superiority and greater applicability of the CESTAC method, the CADNA library, and the novel termination criterion (3.50) over FPA and stopping condition (3.49).

Consider the following linear Volterra integral equation with discontinuous kernel:

$$y(t) = x(t) + \int_0^{\frac{t}{8}} 2ty(\tau)d\tau + \int_{\frac{t}{8}}^{\frac{3t}{8}} (t - \tau)y(\tau)d\tau + \int_{\frac{3t}{8}}^t y(\tau)d\tau,$$

where

$$x(t) = -\cos\left(\frac{3t}{8}\right) + 2\cos(t) - 2t\left(1 - \cos\left(\frac{t}{8}\right) + \sin\left(\frac{t}{8}\right)\right)$$

$$+ \frac{1}{8}\left(-(8 + 7t)\cos\left(\frac{t}{8}\right) + (8 + 5t)\cos\left(\frac{3t}{8}\right)\right)$$

$$- 2\left(-t + (-8 + 5t)\cos\left(\frac{t}{4}\right)\right)\sin\left(\frac{t}{8}\right) + \sin\left(\frac{3t}{8}\right),$$

and the exact solution is $y(t) = \sin t + \cos t$.

In Table 3.2, the numerical results are obtained using the ADM based on FPA for $\varepsilon = 10^{-5}$, and the algorithm is stopped at $n = 6$. Also, in Table 3.3, the numbers of iterations for various ε are shown. It is obvious that for large and small values of ε, accurate results cannot be found. In Table 3.4, the results are obtained based on SA using the CESTAC method and the CADNA library. We do not have ε in this table. The algorithm is stopped at $n_{\text{opt}} = 7$, and it shows the optimal iteration of the ADM for solving this problem. Also, the optimal error is $0.1E-004$, and the optimal approximation is $y_{n_{\text{opt}}} = 0.125085E + 001$.

Table 3.2.   Numerical results for $\varepsilon = 10^{-5}$ based on FPA.

n	$y_{n+1}(t)$	$|y_{n+1}(t) - y(t)|$
1	0.98828512430191040039	0.26257163286209106445
2	1.21838188171386718750	0.03247487545013427734
3	1.24792301654815673828	0.00293374061584472656
4	1.25064921379089355469	0.00020754337310791016
5	1.25084471702575683594	0.00001204013824462891
6	1.25085616111755371094	0.00000059604644775391

Table 3.3.   Number of iterations for different values of ε based on FPA.

ε	Small values	$\varepsilon = 10^{-5}$	$\varepsilon = 10^{-3}$	$\varepsilon = 10^{-1}$	$\varepsilon = 0.5$	Large values
n	$\gg 6$	6	4	2	1	1

Table 3.4.   Numerical results based on the CESTAC method.

n	$y_{n+1}(t)$	$|y_{n+1}(t) - y_n(t)|$	$|y_{n+1}(t) - y(t)|$
1	0.988285E+000	0.988285E+000	0.262571E+000
2	0.121838E+001	0.230096E+000	0.32474E−001
3	0.124792E+001	0.2954E−001	0.293E−002
4	0.125064E+001	0.272E−002	0.20E−003
5	0.125084E+001	0.195E−003	0.1E−004
6	0.125085E+001	0.1E−004	@.0
7	0.125085E+001	@.0	@.0

Let us now consider the following linear Volterra integral equation with discontinuous kernel as the second example:

$$y(t) = x(t) + \int_0^{\frac{t}{9}} (1 + t - \tau)y(\tau)d\tau + \int_{\frac{t}{9}}^{\frac{2t}{9}} y(\tau)d\tau + 2\int_{\frac{2t}{9}}^{\frac{4t}{9}} y(\tau)d\tau$$

$$+ \int_{\frac{4t}{7}}^{t} (t - 1)(\tau + t)y(\tau)d\tau,$$

where

$$x(t) = 2 + \exp\left(\frac{4t}{9}\right) - \exp\left(\frac{t}{9}\right)\left(-1 + \exp\left(\frac{t}{9}\right)\right)$$

$$- 2\exp\left(\frac{2t}{9}\right)\left(-1 + \exp\left(\frac{2t}{9}\right)\right) - \exp\left(\frac{t}{9}\right)\left(2 + \frac{8t}{9}\right) + t,$$

and the exact solution is $y(t) = \exp t$.

The numerical results are obtained based on FPA for $\varepsilon = 10^{-5}$, which are listed in Table 3.5. Also, the numbers of iterations for different values of ε are presented in Table 3.6. The numerical results based on SA are presented in Table 3.7. Using this table, the optimal iteration, the optimal approximation, and the optimal error can be found as $n_{\text{opt}} = 6$, $y_{n_{\text{opt}}} = 0.110516E + 001$, and $E_{n_{\text{opt}}} = 0.8E - 006$, respectively.

Table 3.5. Numerical results for $\varepsilon = 10^{-5}$ based on FPA.

n	$y_{n+1}(t)$	$|y_{n+1}(t) - y(t)|$
1	0.97596895694732666016	0.12920200824737548828
2	1.09789884090423583984	0.00727212429046630859
3	1.10491240024566650391	0.00025856494903564453
4	1.10516428947448730469	0.00000667572021484375

Table 3.6. Number of iterations for different values of ε based on FPA.

ε	Small values	$\varepsilon = 10^{-5}$	$\varepsilon = 10^{-3}$	$\varepsilon = 10^{-1}$	$\varepsilon = 0.5$	Large values
n	$\gg 4$	4	3	1	1	1

Table 3.7. Numerical results based on SA.

n	$y_{n+1}(t)$	$|y_{n+1}(t) - y_n(t)|$	$|y_{n+1}(t) - y(t)|$
1	0.9759688E+000	0.9759688E+000	0.129202E+000
2	0.109789E+001	0.12192E+000	0.7272E−002
3	0.110491E+001	0.7013E−002	0.259E−003
4	0.1105163E+001	0.251E−003	0.7E−005
5	0.1105170E+001	0.6E−005	0.8E−006
6	0.110516E+001	@.0	@.0

Consider the following nonlinear Volterra integral equation with non-smooth kernel:

$$y(t) = x(t) + \int_0^{\frac{t}{2}} (t - \tau) y^2(\tau) d\tau + 2 \int_{\frac{t}{2}}^t y^2(\tau) d\tau,$$

where

$$x(t) = \sin(t) + \frac{1}{16}(2 - 3t^2 - 2\cos(t) + 2t\sin(t))$$
$$+ \frac{1}{2}(-t - \sin(t) + 2\cos(t)\sin(t)),$$

and the exact solution is $y(t) = \sin t$. In Table 3.8, the numerical results are obtained using the CESTAC method and the CADNA library. We can find that the optimal iteration for solving this example using the ADM is $n_{opt} = 6$, the optimal approximation is $y_{n_{opt}} = 0.198821E + 000$, and the optimal error is $0.15E - 004$. The informational zero @.0 shows that the NCSDs between $y_{n+1}(t)$, $y_n(t)$ are almost equal to those between $y_{n+1}(t)$ and $y(t)$. In Table 3.9, the numbers of iterations for different values of ε are obtained based on FPA. We can find that for small values of ε, we have a large number of iterations, while for large values of ε, we do not

Table 3.8.   Numerical results using the CESTAC method and the CADNA library.

n	$y_{n+1}(t)$	$|y_{n+1}(t) - y_n(t)|$	$|y_{n+1}(t) - y(t)|$
1	0.194002E+000	0.194002E+000	0.46670E−002
2	0.198526E+000	0.45239E−002	0.143E−003
3	0.198798E+000	0.272E−003	0.129E−003
4	0.198819E+000	0.20E−004	0.150E−003
5	0.198821E+000	0.1E−005	0.15E−004
6	0.198821E+000	@.0	@.0

Table 3.9.   Number of iterations for different values of ε based on FPA.

ε	Small values	$\varepsilon = 10^{-5}$	$\varepsilon = 10^{-3}$	$\varepsilon = 10^{-1}$	$\varepsilon = 0.5$	Large values
n	$\gg 8$	8	2	1	1	1

have enough iterations. This is one of the main problems with FPA when compared to SA.

Let us consider the following nonlinear Volterra integral equation as the next example:

$$y(t) = x(t) + \int_0^{\frac{t}{7}} t\tau y^3(\tau)d\tau + \int_{\frac{t}{7}}^{\frac{2t}{7}} (t-1)y^3(\tau)d\tau + 3\int_{\frac{2t}{7}}^{\frac{4t}{7}} y^3(\tau)d\tau$$

$$+ \int_{\frac{4t}{7}}^{t} (t-1)(\tau + t)y^3(\tau)d\tau,$$

where

$$x(t) = t^2 - \frac{48641}{5764801}t^7 - \frac{12157553}{46118408}t^8 - \frac{t^9}{46118408},$$

and the exact solution is $y(t) = t^2$. The numerical results of the CESTAC method are presented in Table 3.10. To find these results, we applied the termination criterion (3.50), which depends on two successive approximations. We note that the third column in this table is only for comparison between the results, and we generally do not require the exact solution in the CESTAC method. Based on this table, we can find the optimal iteration as $n_{\text{opt}} = 3$, the optimal approximation as $y_{n_{\text{opt}}} = 0.160000E + 000$, and the optimal error as $E_{n_{\text{opt}}} = 0.1E - 005$. In Table 3.11, we can find the number of iterations of the ADM for solving this example based on FPA.

Finally, the following example is presented to study the sensitivity of $x(t)$ in solving Volterra integral equations with discontinuous kernels.

Table 3.10. Numerical results based on SA.

n	$y_{n+1}(t)$	$|y_{n+1}(t) - y_n(t)|$	$|y_{n+1}(t) - y(t)|$
1	0.159813E+000	0.159813E+000	0.186E−003
2	0.160000E+000	0.187E−003	0.6E−006
3	0.160000E+000	@.0	0.1E−005

Table 3.11. Number of iterations for different values of ε based on FPA.

ε	Small values	$\varepsilon = 10^{-6}$	$\varepsilon = 10^{-3}$	$\varepsilon = 10^{-1}$	$\varepsilon = 0.5$	Large values
n	$\gg 2$	2	1	1	1	1

Consider the following nonlinear Volterra integral equation:

$$y(t) - \int_a^t K(t,\tau)\, y^2(\tau)\, d\tau = x(t), \quad a \le t \le b, \tag{3.63}$$

where the exact solution is

$$y(t) = e^t[\cos(e^t) - e^t \sin(e^t)] \tag{3.64}$$

(which shows fluctuations), and the discontinuous kernel is

$$K(t,\tau) = \begin{cases} 1 - (t-\tau)\, e^{2t}, & a \le \tau \le t/2, \\ p, & \text{otherwise,} \end{cases} \tag{3.65}$$

where p is a parameter.

The direct problem. Let us consider the direct problem of calculating $x(t)$: A nonuniform *grid of nodes*, identical to t and τ, is given by

$$t_1 = \tau_1 = a < t_2 = \tau_2 < \cdots < t_i = \tau_i < \cdots < t_N = \tau_N = b, \tag{3.66}$$

where N is the number of nodes. *The right-hand-side $x(t)$ is calculated* numerically using the trapezoidal formula on grids (3.66) according to the following algorithm:

Algorithm 2

```
p=0.5+1e-10; x(1)=ye(1); %ye is the exact solution (16)
for i=2:N
  int=0;
```

```
  for j=2:i
    int=int+(t(j)-t(j-1))/2*(K(i,j-1)*ye(j-1)^2+K(i,j)
        *ye(j)^2);
  end %j
  x(i)=ye(i)-int;
end %i
```

Here, the following grid of nodes is employed:

$$t = 0(0.1)1.2,\ 1.25(0.05)1.9,\ 1.92(0.02)2.3,\ 2.31(0.01)2.5, \qquad (3.67)$$

i.e. $a = 0$, $b = 2.5$, $N = 67$.

The inverse problem. Let us now consider the inverse problem of solving $y(t)$ of Volterra integral equation:

Algorithm 3 (recurrent solution)

```
p=0.5; y1=x1; h2=t2-t1; h22=h2/2;
y2-h22*(K21*y1^2+K22*y2^2)=x2;
h22*K22*y2^2-y2+x2+h22*K21*y1^2=0; %quadratic equation for y2
y2=(1-\sqrt(1-2*h2*K22*(x2+h22*K21*y1^2)))/(h2*K22); %solution
    of QE
for i=3:N
  yi-\sum_{j=2}^i hj/2*(K(i,j-1)*y(j-1)^2+Kij*yj^2)=xi;
  hi2=hi/2; int=xi+hi2*K(i,i-1)*y(i-1)^2;
  for j=2:i-1
    int=int+hj/2*(K(i,j-1)*y(j-1)^2+Kij*yj^2);
  end %j
  hi2*Kii*yi^2-yi+int=0; %quadratic equation for yi
  yi=(1-\sqrt(1-2*hi*Kii*int))/(hi*Kii); %solution of QE
end %i
```

Figure 3.2 shows the right-hand-side $x(t)$, the exact solution $y(t)$, and the obtained solution $y_n(x)$ (Verlan' and Sizikov, 1986, pp. 41–43). Moreover, the parameter p of the kernel $K(t, \tau)$ in the direct and inverse problems has slightly different values. As a result, the solution at $t \approx 2.5$ (with large fluctuations) differs markedly from the exact solution. Further, regularisation should be applied to increase the stability of the solution.

The CESTAC method is one of the applicable and important methods used to validate numerical results based on SA. To perform validation, instead of the usual applications such as Mathematica, Maple, and

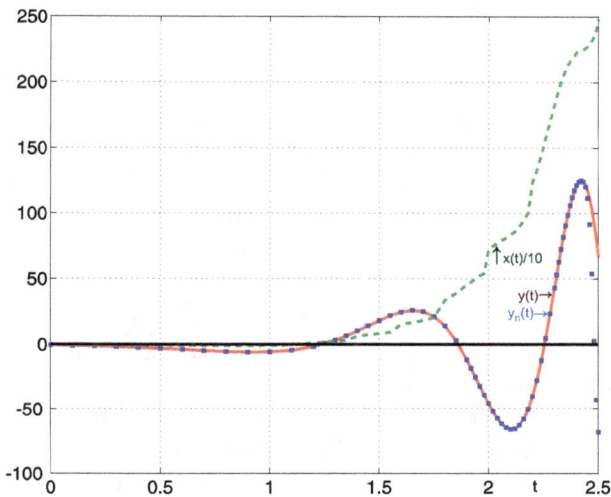

Figure 3.2. Right-hand-side $x(t)$, exact solution $y(t)$, and numerical solution $y_n(t)$.

MATLAB, the CADNA library can be utilised. Using this method and the CADNA library, we can find the optimal iteration, the optimal approximation, and the optimal error of numerical procedures. We introduced the stopping condition based on this method, which is independent of the exact solution. Also, this condition is not limited by the disadvantages of the traditional absolute error condition. Several theorems were proved to show the convergence of the ADM for solving linear and nonlinear Volterra integral equations with discontinuous kernels. The main theorem of the CESTAC method was presented. Based on Theorem 3.4.4, the termination criterion (3.50) can be applied instead of (3.49). Several examples were solved using the ADM, and the numerical results were validated using the CESTAC method. Finally, we compared these results with numerical results obtained using FPA.

3.5   Generalisation of the Frobenius Formula in the Theory of Block Operators on Normed Spaces

The Frobenius formula is a powerful tool in contemporary applications of linear algebra (Gantmacher, 2005). Application of the generalised Frobenius formula for solving singularly perturbed linear equations was considered by Boglaev (1979). Readers may refer to the lectures by Börm *et al.* (2003) on the theory of hierarchical matrices ($\mathcal{H}$-matrices), which makes it possible to approximate certain subblocks of a given matrix by low-rank matrices. For the perturbation theory of linear operators, readers may refer to the studies by Kato (1966), Boichuk and Samoilenko (2016), Jeribi (2015), Albeverio and Elander (2002), Tretter (2008), Sidorov *et al.* (2002), and Lavrentiev and Savelyev (2006). The connection between the Frobenius formula under the construction of generalised inverse operator matrices (also in the sense of Sherman–Morrison–Woodbury formula) and the operator matrices was considered by Malozëmov *et al.* (2002) and Zhuravlev *et al.* (2019). The Frobenius formula is widely used in optimisation, optimal control problems, and high-performance computing (Erofeev *et al.*, 2019).

The construction of approximate methods in bifurcation theory (Sidorov, 1995) in the case of block operators in Banach spaces also employs the Frobenius formula. The latter studies were motivated by a review conducted by Lusternik (1956) and a monograph by Vainberg and Trenogin (1964). The role of skeleton decomposition of linear operators and applications for the solution of irregular systems of partial differential equations are discussed in the work of Sidorov and Sidorov (2017).

Let us now consider real Banach spaces X_j, Y_i, $j, i = 1, 2$ and linear bounded operators $A_{ij} \in \mathcal{L}(X_j \to Y_i)$. Consider the following linear equation:

$$Ax = y, \qquad (3.68)$$

where $A = \begin{bmatrix} A_{11} & A_{12} \\ A_{21} & A_{22} \end{bmatrix}$ is a block operator, $x = (x_1, x_2)^T$, and $y = (y_1, y_2)^T$, with $x_i \in X_i$, $y_i \in Y_i$.

Let us introduce the following conditions:

I. A_{11} is a Fredholm operator (see Vainberg and Trenogin, 1964, Chapters 7 and 9), $\dim N(A_{11}) = n$, $n \geq 0$.

If $n \geq 1$, then $\{\varphi_i\}_1^n$ is the basis in $N(A_{11})$, $\{\gamma_i\}_1^n$ is the corresponding biorthogonal system of functionals from X_1^*, $\{\psi_i\}_1^n$ is the basis in $N(A_{11}^*)$, and $\{z_i\}_1^n$ is the corresponding biorthogonal system of elements from Y_1.

Let us introduce the parameters $\xi_i = \langle x_1, \gamma_i \rangle$ and outline that, for the solvability of the equation $A_{11}x_1 + A_{12}x_2 = y_1$ with respect to x_1, it is necessary and sufficient to fulfil the equalities

$$\langle A_{12}x_2 - y_1, \psi_i \rangle = 0, \quad i = 1, \ldots, n. \qquad (3.69)$$

Here, $\langle \cdot, \cdot \rangle$ denotes the action of a functional on an element in Banach spaces. In Hilbert spaces, $\langle \cdot, \cdot \rangle$ is replaced with $(\cdot, \cdot)$.

Introduce the linear equation

$$\widehat{A}_{11}x_1 + A_{12}x_2 = \sum_{i=1}^{n} \xi_i z_i + y_1, \tag{3.70}$$

where the operator $\widehat{A}_{11} = A_{11} + \sum_{i=1}^{n} \langle \cdot, \gamma_i \rangle z_i$ has a bounded inverse (see Schmidt's Lemma in (Vainberg and Trenogin, 1964, p. 340)). Then, from equation (3.70), x_1 can be expressed as a function of x_2 and the parameters $\xi_1, \ldots, \xi_n$:

$$x_1 = -\widehat{A}_{11}^{-1} A_{12}x_2 + \sum_{i=1}^{n} \xi_i \varphi_i + \widehat{A}_{11}^{-1} y_1. \tag{3.71}$$

Let us substitute x_1 in the second equation of system (3.68) and derive the following equation with respect to x_2:

$$(A_{22} - A_{21}\widehat{A}_{11}^{-1}A_{12})x_2 = y_2 - A_{21}\left(\widehat{A}_{11}^{-1} y_1 + \sum_{i=1}^{n} \xi_i \varphi_i\right). \tag{3.72}$$

II. There exists a bounded operator,

$$L := (A_{22} - A_{21}\widehat{A}_{11}^{-1}A_{12})^{-1} \in \mathcal{L}(Y_2 \to X_2).$$

Remark 3.5.1. If the homogeneous equation $A_{11}x_1 = 0$ has only a trivial solution and A_{11} is a Fredholm operator, then there exists a bounded inverse operator A_{11}^{-1}, and condition **II** becomes

$$(A_{22} - A_{21}A_{11}^{-1}A_{12})^{-1} \in \mathcal{L}(Y_2 \to X_2).$$

This remark holds true for the Frobenius formula, where systems of linear algebraic equations were considered, and $X_1 = Y_1 = \mathbb{R}^m$, $X_2 = Y_2 = \mathbb{R}^{n-m}$, $\mathbb{R}^m$, $\mathbb{R}^{n-m}$ are arithmetic spaces. Obviously, system (3.68), as in the case of the classic Frobenius formula for block matrices, can be rewritten as two equations with two unknown vectors, which are elements of different functional Banach spaces.

The aim of this section is to construct solutions for block equation (3.68) under conditions **I** and **II**. The Frobenius formula for block matrices (Gantmacher, 2005) is generalised in the following.

The remainder of the chapter is organised as follows. First, the main theorems dealing with the solvability of system (3.68) are considered. Asymptotic approximations of solutions are considered next, both in regular and irregular cases. Then, the proposed theory is applied to solving an integro-differential equation. Finally, some concluding remarks are provided. Here, we use the common definitions and conventional notations from the theory of linear operators, e.g. see the textbooks by Trénoguine (1985) and Krantz (2013) and a review by Gokhberg and Krein (1957).

3.5.1   The main theorems on the solvability of a system

We assume conditions **I** and **II** to be fulfilled, and we derive x_2 from equation (3.72):

$$x_2 = L(y_2 - A_{21}\widehat{A}_{11}^{-1}y_1) - \sum_{j=1}^{n}\xi_j LA_{21}\varphi_j. \qquad (3.73)$$

In the case of $\dim N(A_{11}) = n$, it remains to determine the parameters $\xi_1, \ldots, \xi_n$. These parameters can be found from the following system of linear algebraic equations constructed through substitution of x_2 by virtue of the solvability condition (3.69):

$$B\xi = d, \qquad (3.74)$$

where

$$B = [\langle A_{12}LA_{21}\varphi_j, \psi_i\rangle]_{i,j=1}^{n}$$

is an $n \times n$ matrix and

$$d = \langle A_{12}L(y_2 - A_{21}\widehat{A}_{11}^{-1}y_1) - y_1, \psi_i\rangle\big|_{i=1}^{n}$$

is a vector in $\mathbb{R}^n$. Therefore, the following statements are evident.

Theorem 3.5.2. *Let conditions* **I** *and* **II** *be fulfilled. Then, for the unique solvability of system (3.68) with an arbitrary right-hand side, it is necessary and sufficient for the matrix B to be nondegenerate.*

Corollary 3.5.3. *Let the conditions in Theorem 3.5.2 be fulfilled and $\det B \neq 0$, then the operator A is continuously invertible, $\xi = B^{-1}d$, and using formulae (3.71) and (3.73), the operator A^{-1} can be explicitly constructed in block form.*

Theorem 3.5.4. *Let conditions* **I** *and* **II** *be fulfilled, rank $B = r$, and $B^*c_i = 0$, $i = 1, \ldots, n - r$. If $\sum_{i=1}^{n-r}|(d, c_i)| = 0$, then system (3.68) has an $n - r$-parameter family of solutions.*
If $\sum_{i=1}^{n-r}|(d, c_i)| \neq 0$, then system (3.68) has no solutions.

The proof of Theorem 3.5.2 follows from (3.71), (3.73) and the known conditions for the solvability of systems of linear algebraic equations.

From Theorem 3.5.2, the generalisation of the classic Frobenius formula follows in the case of block operators in normed spaces.

Corollary 3.5.5. *Let the operators A_{11} and $(A_{22} - A_{21}A_{11}^{-1}A_{12})$ be continuously invertible. Then, system (3.68) has a unique solution, and the operator A^{-1} can be constructed explicitly in Frobenius block form.*

Proof. Indeed, in that case, $x_2 = (A_{22} - A_{21}A_{11}^{-1}A_{12})^{-1}(y_2 - A_{21}A_{11}^{-1}y_1)$ and $x_1 = -A_{11}^{-1}A_{12}x_2 + A_{11}^{-1}y_1$. These formulae allow us to represent the inverse operator A^{-1} in block form. □

Furthermore, due to the conditions imposed in the corollary, the solution of the given system is unique and defined as

$$x = A^{-1}y,$$

where

$$A^{-1} = \begin{bmatrix} M_{11} & M_{12} \\ M_{21} & M_{22} \end{bmatrix},$$

with $M_{22} = (A_{22} - A_{21}A_{11}^{-1}A_{12})^{-1}$, $M_{21} = -M_{22}A_{21}A_{11}^{-1}$, $M_{11} = A_{11}^{-1} + A_{11}^{-1}A_{12}M_{22}A_{21}A_{11}^{-1}$, $M_{12} = -A_{11}^{-1}A_{12}M_{22}$, and

$$M_{ij} \in \mathcal{L}(Y_j \to X_i).$$

The block operator A^{-1} is a generalisation of the Frobenius formula (Gantmacher, 2005) in linear normed spaces.

The following corollary can be verified in a similar way.

Corollary 3.5.6. *If A_{11} is a Fredholm operator, $\dim N(A_{11}) = n$, $\det B \neq 0$, then there exists an inverse operator, A^{-1}, that can be explicitly constructed.*

Indeed, due to the conditions imposed in the corollary, the solution of the given system is unique and defined as

$$x_2 = L(y_2 - A_{21}\widehat{A}_{11}^{-1} - A_{21}(\xi, \phi)),$$
$$x_1 = \widehat{A}_{11}^{-1}(y_1 - A_{21}x_x) + (\xi, \phi).$$

Here, $(\xi, \phi) = \sum_1^n \xi_i \phi_i$, $\xi = B^{-1}d$. The operators L and $\widehat{A}_{11}^{-1}$, the matrix B, and the vector d (which is linear in y_1 and y_2) are defined above. The constructed solution, similar to the result of the corollary, can be rewritten in block form as $x = A^{-1}y$. This result reinforces the Frobenius approach to handling block matrices.

3.5.2   *Asymptotic approximations of solutions*

In this section, we assume that the norm of one of the operators A_{21}, A_{12} is sufficiently small.

Regular case.   Let the following condition be fulfilled:

III. Assume that each of the operators A_{11}, A_{22} has a bounded inverse, $\min(||A_{12}||, ||A_{21}||) \le \delta$. Then, without loss of generality, we assume that $||A_{12}|| \le \delta$ and consider the following system:

$$A_{11}x_1 + \lambda A_{12}x_2 = y_1, \quad A_{21}x_1 + A_{22}x_2 = y_2 \qquad (3.75)$$

with a small parameter, λ. For sufficiently small $\delta > 0$, there exists a unique solution for this system, which can be presented as a series expansion in powers of the small parameter λ.

Indeed, through the substitution of the series $x_1 = \sum_{i=0}^{\infty} a_i \lambda^i$, $x_2 = \sum_{i=0}^{\infty} b_i \lambda^i$ in system (3.75), taking into account the invertibility of the operators A_{11}, A_{22}, the coefficients a_i, b_i can be determined using the undefined coefficients method from the following equalities:

$$\sum_{i=0}^{\infty} (a_i + \lambda A_{11}^{-1} A_{12} b_i) \lambda^i = A_{11}^{-1} y_1,$$

$$\sum_{i=0}^{\infty} (b_i + A_{22}^{-1} A_{21} a_i) \lambda^i = A_{22}^{-1} y_2.$$

Therefore, $a_0 = A_{11}^{-1} y_1$, $b_0 = A_{22}^{-1} y_2 - A_{22}^{-1} a_{21} a_0$, $a_i = -A_{11}^{-1} A_{12} b_{i-1}$, $b_i = -A_{22}^{-1} a_{21} a_i$, $i = 1, 2, \ldots$ Series convergence in the small neighbourhood $|\lambda| \le \rho$ follows from the well-known inverse operator theorem (see Trénoguine, 1985, pp. 141–142). It is easy to estimate the range and speed of convergence. Indeed,

$$x_2 = A_{22}^{-1}(y_2 - A_{21} x_1),$$

where x_1 is the solution of the equation

$$(A_{11} - \lambda A_{12} A_{22}^{-1} A_{21}) x_1 = y_1 - \lambda A_{12} A_{22}^{-1} y_1.$$

We fix $q < 1$ to be positive and

$$\rho = q ||A_{11}^{-1} A_{12} A_{22}^{-1} A_{21}||^{-1}.$$

Then, for $|\lambda| \leq \rho$, the element x_1 can be constructed using the formula

$$x_1 = \sum_{n=0}^{\infty} \lambda^n (A_{11}^{-1} A_{12} A_{22}^{-1} A_{21})^n A_{11}^{-1} (y_1 - \lambda A_{12} A_{22}^{-1} y_1).$$

Since X_1 is a Banach space and $q < 1$, the series converges for $|\lambda| \leq \rho$ at the rate of a geometrical progression with denominator q.

Then, the following result is valid.

Theorem 3.5.7. *Assume that each of the operators A_{11}, A_{22} has a bounded inverse and one of the operators A_{21}, A_{12} has a sufficiently small linear operator norm. Then, the operator A has a bounded inverse, and moreover, system (3.75) has a unique analytical solution in the neighbourhood $|\lambda| \leq \rho$.*

Irregular case. Let the following condition be fulfilled:

IV. A_{11} *is a Fredholm operator,* $\dim N(A_{11}) = 1$, *and A_{22} is an invertible operator.*

For the sake of clarity, let us consider that calculations for this type of system can be presented as system (3.75) under the condition of $\dim N(A_{11}) = 1$.

We prove that under condition **IV**, a solution to system (3.75) can be represented as a sum of a Laurent series with a pole of order 1 at the origin.

Then, we employ the following ansatz:

$$x_1 \sim \frac{a_{-1}}{\lambda} + \sum_{i=0}^{\infty} a_i \lambda^i,$$

$$x_2 \sim \frac{b_{-1}}{\lambda} + \sum_{i=0}^{\infty} b_i \lambda^i.$$

The coefficients a_i, b_i can be found using the undetermined coefficients method, taking into account the Nekrasov–Nazarov method (Akhmedov, 1957).

Indeed, from the equalities

$$\sum_{i=0}^{\infty} (A_{11} a_{i-1} + \lambda A_{12} b_{i-1}) \lambda^{i-1} = y_1,$$

$$\sum_{i=0}^{\infty} (b_{i-1} + A_{22}^{-1} A_{21} a_{i-1}) \lambda^{i-1} = A_{22}^{-1} y_2,$$

the following equations can be derived:

$$A_{11}a_{-1} = 0,$$

$$b_{-1} = -A_{22}^{-1}A_{21}a_{-1},$$

$$A_{11}a_0 + A_{12}b_{-1} = y_1,$$

$$b_0 = -A_{22}^{-1}A_{21}a_0 + A_{22}^{-1}y_2,$$

$$A_{11}a_i + A_{12}b_{i-1} = 0,$$

$$b_i = -A_{22}^{-1}A_{21}a_i, \quad i = 1, 2, \ldots.$$

Due to the Fredholm property of the operator A_{11}, when one solves these equations, it is necessary to take the following conditions into account: $\langle y_1 - A_{12}b_{-1}, \psi \rangle = 0$, $\langle A_{12}b_{i-1}, \psi \rangle = 0$, $i = 1, 2, \ldots$.

Therefore, $a_{-1} = c_{-1}\varphi$, $b_{-1} = -A_{21}^{-1}A_{21}c_{-1}\varphi$. The constant value c_{-1} can be defined from the linear algebraic equality

$$\langle y_1 + A_{12}A_{22}^{-1}A_{21}c_{-1}\varphi, \psi \rangle = 0.$$

Therefore, we require

$$\langle A_{12}A_{22}^{-1}A_{21}\varphi, \phi \rangle \neq 0.$$

Then, $a_0 = c_0\varphi + \widehat{a}_0$, $b_0 = A_{21}^{-1}y_2 - A_{22}^{-1}A_{21}(c_0\varphi + \widehat{a}_0)$. The constant value c_0 can be defined using the equality $\langle A_{12}b_0, \psi \rangle = 0$. The rest of the coefficients of the Laurent series can be defined in a similar way. Thus, the following theorem is true.

Theorem 3.5.8. *Let A_{11} be a Fredholm operator, $\dim N(A_{11}) = 1$, $A_{11}\varphi = 0$, $A_{11}^*\psi = 0$, and let operator A_{22} be invertible. Assume that*

$$\langle A_{12}A_{22}^{-1}A_{21}\varphi, \psi \rangle \neq 0.$$

Then, system (3.75) in the neighbourhood $0 < |\lambda| < \rho$ has a solution that can be represented as a Laurent series with a first-order pole.

Remark 3.5.9. *If $\dim N(A_{11}) = n$, $n > 1$, $\{\varphi_i\}_1^n$ is a basis in $N(A_{11})$, $\{\psi_i\}_1^n$ is a basis in the adjoint $N(A_{11}^*)$, and $\det\langle A_{12}A_{22}^{-1}A_{21}\phi_i, \psi_j \rangle_{i,j=1}^n \neq 0$, then the result of the theorem is true.*

The proof of this remark follows the same steps as those listed in the above proof for the case of $n = 1$. Indeed, we seek a solution in the form of a Laurent series with a first-order pole. The coefficients of the series are determined, as before, using the Nekrasov–Nazarov method. According to this method, at each step, n arbitrary constants can be determined from n conditions of solvability of the subsequent equations of the recurrent chain of equations.

Remark 3.5.10. If $\det\langle A_{12}A_{22}^{-1}A_{21}\phi_i, \psi_j\rangle_{i,j=1}^n = 0$, then under additional conditions (see Vainberg and Trenogin, 1964, Section 31), a solution for system (3.75) can be constructed as a Laurent series with a pole of order $p > 1$.

Remark 3.5.11. The calculation of the coefficients of the Laurent series is carried out for system (3.68) reduced to the form (3.75). Obviously, in the case of the reduction of system (3.68) to

$$A_{11}x_1 + A_{12}x_2 = y_1, \quad \lambda A_{21}x_1 + A_{22}x_2 = y_2, \tag{3.76}$$

under condition of Theorem 3.5.8, the coefficients of the Laurent series are calculated in a similar way.

Let us consider the following the boundary value problem (BVP).

Example 3.5.1.

$$\begin{cases} \frac{\partial^2 u_1(x,t)}{\partial x^2} + au_1(x,t) + u_2(x,t) = y_1(x,t), \\ u_1(0,t) = 0, \ u_1(\pi,t) = 0, \\ \lambda \int_0^\pi K(x,s)u_1(s,t)\, ds + b(x)u_2(x,t) = y_2(x,t), \end{cases} \tag{3.77}$$

where $y_1(x,t)$, $y_2(x,t)$, $K(x,s)$, $b(x)$ are known continuous functions for $x \in [0,\pi]$ and $t \in [0,\infty]$, $y_i(0,t) = y_i(\pi,t) = 0$, $b(x) \neq 0$ for $x \in [0,\pi]$, λ is a small parameter.

Let us construct the classic continuous solution of BVP (3.77). Let us introduce the operators $(A_{11}u_1)(x,t) = \frac{\partial^2 u_1(x,t)}{\partial x^2} + au_1(x,t)$, $(A_{12}u_2)(x,t) = u_2(x,t)$, $(A_{21}u_1)(x,t) = \lambda \int_0^\pi K(x,s)u_1(s,t)\, ds$, $(A_{22}u_2)(x,t) = b(x)u_2(x,t)$. Then, system (3.77) can be represented in form of (3.68).

There are two possible cases: regular $(a \neq n^2)$ and irregular $(a = n^2)$, where $n \in \mathbb{N}^+$.

In the *regular case*, by employing Theorem 3.5.7, the solutions can be constructed as the following convergent series in the neighbourhood of the point $\lambda = 0$:

$$u_1(x,t) = \sum_{i=0}^{\infty} a_i(x,t)\lambda^i,$$

$$u_2(x,t) = \sum_{i=0}^{\infty} b_i(x,t)\lambda^i.$$

In the *irregular case* $a = n^2$, the homogeneous boundary problem

$$\begin{cases} \frac{\partial^2 u_1}{\partial x^2} + n^2 u_1 = 0, \\ u_1(0,t) = u_1(\pi,t) = 0 \end{cases} \tag{3.78}$$

has the nontrivial solution $c(t)\sin nx$, where $c(t)$ is an arbitrary function. Therefore, for solution construction, we employ the Laurent series (see Theorem 3.5.8).

The inequality in Theorem 3.5.8 is fulfilled if

$$\int_0^\pi \int_0^\pi \frac{K(x,s)}{b(x)} \sin ns \, \sin nx \, dx ds \neq 0.$$

Since

$$u_2(x,t) = \frac{1}{b(x)} \left[y_2(x,t) - \lambda \int_0^\pi K(x,s)u_1(s,t)\, ds \right],$$

for solution construction, it is sufficient to construct the solution of the boundary problem

$$\begin{cases} \frac{\partial^2 u_1(x,t)}{\partial x^2} + n^2 u_1(x,t) - \lambda \int_0^\pi \frac{K(x,s)}{b(x)} u_1(s,t)\, ds = y_1(x,t) - \frac{y_2(x,t)}{b(x)}, \\ u_1(0,t) = u_1(\pi,t) = 0 \end{cases}$$

$$\tag{3.79}$$

as the series

$$u_1 = \sum_{i=-1}^{\infty} a_i(x,t)\lambda^i.$$

The coefficients $a_i(x,t)$, $i = -1,0,1,2,\ldots$ can be defined as shown in the proof of Theorem 3.5.8 using the Nekrasov–Nazarov method (Akhmedov, 1957), which adopts a two-step method of undetermined coefficients. In this case, the functions of the argument t are calculated from the conditions of solvability of the subsequent equations of the recurrent chain of linear equations.

Indeed,

$$
\begin{cases}
\frac{\partial^2 a_{-1}}{\partial x^2} + n^2 a_{-1} = 0, \\
a_{-1}|_{x=0} = a_{-1}|_{x=\pi} = 0,
\end{cases}
\tag{3.80}
$$

$$
\begin{cases}
\frac{\partial^2 a_0}{\partial x^2} + n^2 a_0 = \int_0^\pi \frac{K(x,s)}{b(x)} a_{-1}(s,t)\, ds + y_1(x,t) - \frac{y_2(x,t)}{b(x)}, \\
a_0|_{x=0} = a_0|_{x=\pi} = 0.
\end{cases}
\tag{3.81}
$$

Then, $a_{-1}(x,t) = c_{-1}(t)\sin nx$, where the function $c_{-1}(t)$ can be determined from the conditions of solvability of system (3.81), i.e. from equation

$$
c_{-1}(t) \int_0^\pi \int_0^\pi \frac{\sin nx K(x,s)}{b(x)} \sin ns \, dx ds
$$
$$
+ \int_0^\pi \sin nx \left(y_1(x,t) - \frac{y_2(x,t)}{b(x)} \right) dx = 0.
$$

The solution of boundary problem (3.81) is $a_0(x,t) = c_0(t)\sin nx + \hat{a}_0(x,t)$, where $c_0(t)$ is an arbitrary function and $\hat{a}_c(x,t)$ is the particular solution of boundary problem (3.81). The function $c_0(t)$ can be similarly defined from the conditions for the solvability of the BVP with respect to the following coefficient $a_1(x,t)$ of the Laurent series.

Remark 3.5.12. Results close to Theorem 3.5.2 can be found in the work of Zhuravlev *et al.* (2019). The method proposed by Zhuravlev *et al.* (2019) requires the prior construction of special projectors and operator constriction. However, the explicit form of the block inverse matrix in the paper is not specified. In our work, the inverse operator is constructed in closed form. A constructive role in our approach is played by the Schmidt–Trenogin pseudo-resolvent operator $\widehat{A}_{11}^{-1}$, which has regularising properties in the Tikhonov–Lavrentiev sense, thereby providing stability of calculations (see Trénoguine, 1985, Section 22, pp. 233–237).

In the general case, when solving block system (3.68) with a normally solvable operator A_{11}, one can use the Laurent series with a higher-order pole. Indeed, if the operator A_{22} is invertible, then the problem can be reduced to that of solving the linear equation $A_{11}x_1 - Bx_1 = y_1 - A_{12}A_{22}^{-1}y_1$, where $B = A_{12}A_{22}^{-1}A_{21}$. If the norm of one of the operators A_{12}, A_{21} is sufficiently small, the equation can be written as $A_{11}x_1 - \lambda B x_1 = y_1 - A_{12}A_{22}^{-1}y_2$, where $B \in \mathcal{L}(Y_1 \to X_1)$, λ is a small parameter.

If the normally solvable operator A_{11} has a complete B-Jordan set and the length of the maximal B-Jordan chain is equal to p, then the solution can be constructed in the form of a Laurent series with a pole at the point $\lambda = 0$ of order p based on the known facts of perturbation theory. If the value region of the operator A_{11} is not closed (i.e. the operator is not normally solvable), then we can use regularisation methods (Lavrentiev and Savelyev, 2006) in the sense of Tikhonov and Lavrentiev to construct solutions for the equations. The approach studied in this section can be employed to analyse complex systems with various functional operators.

Chapter 4

Loaded Equations and Bifurcation Analysis

4.1 Nonlinear Loaded Volterra Equations of the First Kind with Bifurcation Parameters: Existence Theorem and Solution Construction

Let us consider the nonlinear Volterra integral equation with local and integral (i.e. non-local) loads on the sought solution. The loads are given by means of Stieltjes integrals. The equation contains a bifurcation parameter, at any value of which the equation has a trivial solution. We derive the necessary and sufficient conditions on the values of the parameter (bifurcation points) in the neighbourhood of which the equation has a nontrivial bifurcation solution. The principal terms of the asymptotic of the branch solutions are constructed, and the method for their refinement is formulated. We then provide practical examples to support the proposed theory.

Let us consider the equation

$$x(t) = \mathcal{L}(x(t), x_\alpha, t, \lambda), \tag{4.1}$$

where $t \in [0, T]$, the desired function is $x(t) \in C_{[0,T]}$, $\lambda \in \mathbb{R}^1$ is a bifurcation parameter, and x_α is given by the linear Stieltjes functional $\int_{t_1}^{t_2} x(t)\, d\alpha(t)$, where $\alpha(t)$ is a bounded function of variation, $[t_1, t_2] \in (0, T)$. Therefore, the load x_α can be either local of the form $x_\alpha = x(\alpha)$, $\alpha \in (0, T)$, or non-local. The nonlinear integral operator $\mathcal{L}(x(t), x_\alpha, t, \lambda)$ is defined by formula (4.2).

The equations with loadings are similar to non-standard differential-integral equations (Sidorov *et al.*, 2020c) and neutral-type equations (Akhmetov *et al.*, 1982) in properties and in the fields of application.

Note that the method of convex majorants allows us to investigate the existence regions of nonlinear integral equations (cf. Sidorov and Sidorov, 2012).

In recent years, certain parts of the theory of differential and integral equations have developed into independent branches. However, even the simplest classes of Volterra equations with loadings and bifurcation parameters have remained unexplored until recently. The aim of this section is to summarise the results in this field as a rather general formulation.

Let us first focus on the problem statement. Let us consider the following condition:

I. The nonlinear mapping is of the form

$$\mathcal{L}(x(t), x_\alpha, t, \lambda) = a(t, \lambda)x_\alpha + \sum_{i+k=l}^{N} f_{ik}(t, \lambda)x^i(t)x_\alpha^k$$

$$+ \int_0^t \sum_{i+k=l}^{N} K_{ik}(t, s\lambda)x^i(s)x_\alpha^k \, ds + R(x(t), x_\alpha, t, \lambda),$$

$$(4.2)$$

where $l \geq 2$, $\lambda \in \mathbb{R}^1$, $t \in [0, T]$,

$$|R(x(t), x_\alpha, t, \lambda)| = \mathcal{O}(|x(t)| + |x_\alpha|^{N+1}).$$

The functions $a(t, \lambda)$, $f_{ik}(t, \lambda)$, and $K_{ik}(t, s, \lambda)$ are continuous and sufficiently smooth on the parameter λ in the neighbourhood of the bifurcation point λ_0. Obviously, for any λ, equation (4.1) is satisfied by the pair $x(t) = 0$ and $x_\alpha = 0$.

Definition 4.1.1. The point λ_0 is called the bifurcation point of equation (4.1) if, for any $\varepsilon > 0$, $\delta > 0$, there are $x(t)$ and λ satisfying (4.1) and such that $0 < ||x|| < \varepsilon$, $|\lambda - \lambda_0| < \delta$.

In this section, we find the conditions under which the point $\lambda_0 \in \mathbb{R}^1$ is a bifurcation point of equation (4.1), and one can then construct the asymptotics of the nontrivial branches of the small solutions of equation (4.1).

A branch is a continuous real solution $x(t)$ such that $x(t) \to 0$ at $\lambda \to \lambda_0 + 0$ (or $\lambda \to \lambda_0 - 0$). Obviously, each solution $x(t)$ is assigned a single load by a given linear functional.

In a number of works, it was noted that the loaded equations model processes whose course at certain moments (or local time intervals) influences their course as a whole. A unique difficulty in the study of such problems is caused by the practical impossibility of making measurements at these moments. Many studies have been devoted to loaded differential equations (ordinary and partial). However, classes of Volterra integral equations with loads and bifurcation analysis in models with loads have not been studied in the literature before.

We obtain a solution to the problem by constructing an equation with respect to the load with a bifurcation parameter λ and investigating it using the method of successive approximations, the method of degree geometry, and rotations of finite-dimensional vector fields.

4.1.1   *Building the equation with respect to load and necessary bifurcation conditions*

For the sake of clarity, we first give an exposition under a single load. Using the method of successive approximations and the implicit representation theorem in the analytic case, the solution $x(t)$ of the Volterra integral equation (4.1) at $t \in [0, T]$, for $|\lambda - \lambda_0| \leq \rho_1$ and $|x_\alpha| \leq \rho_2$ is the series

$$x(t) = \sum_{n=1}^{\infty} a_n(t, \lambda) x_\alpha^n. \tag{4.3}$$

Let us sequentially calculate the functions $a_n(t, \lambda)$ using the recurrence formulae

$$a_1(t, \lambda) = a(t, \lambda),$$

$$a_n(t, \lambda) = \frac{1}{n!} \frac{\partial^n}{\partial x_\alpha^n} \mathcal{L} \left(\sum_{k=1}^{n-1} a_k(t, \lambda) x_\alpha^k, x_\alpha, t, \lambda \right) \Bigg|_{x_\alpha = 0}, \quad n = 2, 3, \dots . \tag{4.4}$$

In this case, due to the structure of operator (4.2), we can easily obtain simple formulas for the first $(2l - 2)$ coefficients of series (4.3): $a_2(t, \lambda) = \dots = a_{l-1}(t, \lambda) \equiv 0$,

$$a_n(t, \lambda) = \frac{1}{n!} \sum_{i+k=n} \left[f_{ik}(t, \lambda) a^i(t, \lambda) + \int_0^t K_{ik}(t, s, \lambda) a^i(s, \lambda) \, ds \right],$$

$n = l, l+1, \dots, 2l - 2.$

Based on the implicit mapping theorem, series (4.3) will converge in a sufficiently small neighbourhood of $|x_\alpha| \leq \rho_2$. The value of ρ_2

can be estimated, as demonstrated in the following, using the convex majorants method (Sidorov and Sidorov, 2012; Sidorov, 2014a). From the aforementioned, we have the following lemma.

Lemma 4.1.1. *Let condition* **I** *be satisfied. Then, the load* x_α *satisfies the equation*

$$A_1(\lambda)x_\alpha + \sum_{i=l}^{\infty} A_i(\lambda)x_\alpha^i = 0, \qquad (4.5)$$

where

$$A_1(\lambda) = \int_{t_1}^{t_2} a_1(t, \lambda)\, d\alpha(t) - 1,$$

$$A_i(\lambda) = \int_{t_1}^{t_2} a_i(t, \lambda)\, d\alpha(t), \quad i = l, l+1, \dots.$$

Proof. The proof follows from the possibility of representing the solution $x(t)$ in the form of a series (4.3) and setting the load by means of a linear functional. □

Corollary 4.1.2 (necessary conditions for bifurcation). *For λ_0 to be the bifurcation point, it is necessary that $A_1(\lambda_0) = 0$.*

Proof. Equation (4.5) at λ has a trivial solution, $x_\alpha \equiv 0$. If $A_1(\lambda_0) \neq 0$, then at $|\lambda - \lambda_0| \leq \rho_1$ in the neighbourhood of $|x_\alpha| < \rho_2$. On the basis of the implicit function theorem, the small solution of equation (4.5) is singular. Hence, in this case, $x_\alpha \equiv 0$, and λ_0 is not a bifurcation point according to Definition 4.1.1 of equation (4.1). □

Corollary 4.1.3. *Suppose that in equation (4.5) all the coefficients of $A_i(\lambda)$ at the point λ_0 are zero. Then, λ_0 will be the bifurcation point. Moreover, equation (4.1) at $\lambda = \lambda_0$ has a c-parametric nontrivial $x(t, c)$, depending on a sufficiently small value of the parameter c. For $0 < |\lambda - \lambda_0| < \rho_1$, equation (4.1) has no other small solutions.*

Proof. The proof is obvious since, by the conditions of Corollary 4.1.3, the load x_α at $\lambda = \lambda_0$ in the expansion (4.3) of the solution of equation (4.1) remains an arbitrary parameter c from the interval $|c| \leq \rho_2$, to which series (4.3) converges. □

4.1.2 Sufficient conditions for the existence of a bifurcation point and asymptotics of solutions

Constructive sufficient conditions for the existence of bifurcation points are obtained by defining real solutions, $x_\alpha \to 0$, at $\lambda \to \lambda_0 + 0$ ($\lambda \to \lambda_0 - 0$) of equation (4.5) and substituting them into the right-hand side of formula (4.3). Recall that, in equation (4.5),

$$A_l(\lambda) = \frac{1}{l!} \sum_{i+k=l} \sum \left(f_{ik}(\alpha, \lambda) a^i(\alpha, \lambda) + \int_0^\alpha K_{ik}(\alpha, s, \lambda) a^i(s, \lambda) \, ds \right)$$

in the case of a local load and

$$A_l(\lambda) = \frac{1}{l!} \int_{t_1}^{t_2} \sum_{i+k=l} [f_{ik}(t, \lambda) a^i(t, \lambda) + \int_0^t K_{ik}(t, s, \lambda) a^i(s, \lambda) \, ds] \, d\alpha(t)$$

in the case of an integral (non-local) load.

By virtue of Lemma 4.2.1, to construct the asymptotics of the function x_α following equation (4.5), put $\lambda = \lambda_0 + \mu$, where μ is a small real parameter. Let's introduce the following conditions:

II. λ_0, the root of the equation $A_1(\lambda) = 0$ of multiplicity p;
III. $A_l(\lambda_0) \neq 0$.

Then, equation (4.5) with $\lambda = \lambda_0 + \mu$ is transformed into the form

$$\left(\frac{1}{p!} A_1^{(p)}(\lambda_0) \mu^p + \mathcal{O}(|\mu|^{p+1}) \right) x_\alpha + (A_l(\lambda_0) + \mathcal{O}(|\mu|)) x_\alpha^l + \mathcal{O}(|x_\alpha|^{l+1}) = 0 \tag{4.6}$$

in the neighbourhood of points $x_\alpha = 0$ and $\mu = 0$. We look for a solution to equation (4.6) using the Newton–Puiseux series (with a Newton's diagram used to determine the exponent $p/(l-1)$):

$$x_\alpha = (c_0 + \mathcal{O}(|\mu|)) \mu^{\frac{p}{l-1}}, \quad c_0 \neq 0.$$

To determine c_0, we obtain the equation

$$\operatorname{sign} \mu^p \frac{1}{p!} \left. \frac{d^p A_1(\lambda)}{d\lambda^p} \right|_{\lambda=\lambda_0} c + A_l(\lambda_0) c^l = 0. \tag{4.7}$$

Therefore, it is obvious that for odd p, there are two equations, namely,

$$\frac{1}{p!} A_1^{(p)}(\lambda_0) c + A_l(\lambda_0) c^l = 0, \quad \text{for } \mu > 0, \tag{4.8}$$

$$-\frac{1}{p!} A_1^{(p)}(\lambda_0) c + A_l(\lambda_0) c^l = 0, \quad \text{for } \mu < 0. \tag{4.9}$$

For odd p and any l in equation (4.7), at least one semi-neighbourhood point $\mu = 0$ has a simple real solution, $c_0 \neq 0$.

Case 1.
If l is even, equation (4.8) will satisfy the real solution

$$c_0 = \sqrt[l-1]{-\frac{1}{p!}\frac{A_1^{(p)}(\lambda_0)}{A_l(\lambda_0)}},$$

and equation (4.6) will have a real solution, $x_\alpha \sim c_0|\mu|^{p/(l-1)}$, for $\mu \to +0$, and accordingly, $x_\alpha \sim -c_0|\mu|^{p/(l-1)}$ at $\mu \to -0$.

Case 2.
For odd l and $A_1^{(p)}(\lambda_0)A_l(\lambda_0) < 0$, equation (4.8) has two real solutions,

$$c_{0\pm} = \pm\sqrt[l-1]{\frac{1}{p!}\frac{A_1^{(p)}(\lambda_0)}{A_l(\lambda_0)}},$$

and equation (4.6) in the semi-neighbourhood of $0 \leq \mu \leq \rho_1$ enjoys two real solutions, $x_{\alpha\pm} \sim c_{0\pm}\mu^{p/(l-1)}$. For odd l and $A_1^{(p)}(\lambda_0)A_l(\lambda_0) > 0$, we get a similar result for $-\rho_1 \leq \mu \leq 0$.

From the above, we get the following theorem.

Theorem 4.1.4. *Let conditions* **I, II,** *and* **III** *be satisfied. Then, λ_0 is the bifurcation point of equation* (4.1). *Moreover, for even l, equation* (4.1) *has the only small real solution $x(t, \lambda)$ in the neighbourhood of the point λ_0:*

$$x(t,\lambda) \sim a_1(t,\lambda_0)\sqrt[l-1]{-\frac{1}{p!}\frac{A_1^{(p)}(\lambda_0)}{A_l(\lambda_0)}}(\lambda - \lambda_0)^{p/(l-1)}.$$

For odd l, equation (4.1) *has two real solutions with similar asymptotics in one of the semi-neighbourhoods of the point λ_0.*

From Theorem 4.1.4, taking into account Corollary 4.1.3, we have the following theorem.

Theorem 4.1.5. *Let conditions* **I** *and* **II** *be satisfied. Let λ_0 be the root of the equation $A_{11}(\lambda) = 0$ of odd multiplicity. Then, λ_0 is the bifurcation point of equation* (4.1).

Theorem 4.1.4 takes into account the possibility of representing other solution varieties of equation (4.5) in the form of Newton–Puiseux expansions to allow for generalisation.

Indeed, consider the first few coefficients, $A_1(\lambda)$, $A_l(\lambda)$, $A_{l+1}(\lambda), \ldots,$ $A_{n-1}(\lambda)$, of equations (4.5).

Let's introduce the following condition:

IV. The number λ_0 is a zero of the corresponding multiplicities p_i, $i = 1, l, \ldots, n-1$ of above coefficients, and $A_n(\lambda_0) \neq 0$.

Let's plot the points $(1, p_1)$, (l, p_2), $\ldots, (n-1, p_{n-1}), (n, 0)$ on the coordinate plane and construct a Newton's diagram of a set consisting of these points.

Take the face (a, b). This face contains the points (i, p_i), satisfying the equality $ir/s + k = \theta$, where $r/s = \tan\varphi$ and θ is the ordinate of the intersection point of the continuation of the face (a, b) with the y-axis. Obviously, θ will also be a rational number.

We look for a solution to equation (4.5) in the form

$$x_\alpha = c(\mu)|\mu|^{r/s}, \quad c(0) \neq 0.$$

Then, $c(0)$ must satisfy at least one of the two equations

$$P_\pm(c) := \sum_{ir/s+k=\theta} \operatorname{sign}\mu^k \frac{1}{p_i!} a_i^{(p_i)}(\lambda_0)c^i = 0.$$

Having constructed simple real solutions, $c^* \neq 0$, for these equations, we can, as in the case of Theorem 4.1.4, write out the asymptotics of small solutions, $x(t, \lambda)$, of equation (4.1) in two semi-neighbourhoods of the bifurcation point λ_0.

4.1.3 Application of the method of undetermined coefficients

The above approach allows us to construct solutions to equation (4.1) directly as a series in the integer or fractional power of the parameter $\lambda - \lambda_0$, where λ_0 is the bifurcation point. Coefficients of the series, as in the Nekrasov–Nazarov method (Akhmedov, 1957), will be determined from the recurrent sequence linear equations. The peculiarity of this analytical method is that arbitrary constants at each step are determined from the solvability conditions following sequence equations.

The following is an example demonstrating this method of solving equations with loads under the conditions of Theorem 4.1.4.

Let us consider the equation

$$x(t) = t(\lambda - 2)x(t) - \int_0^t tx^3(s)\, ds.$$

For equation (4.5) with a load, the following coefficients can be obtained: $A_1(\lambda) = \lambda - 3$, and $A_3(\lambda) = \frac{-1}{5}(\lambda - 2)^3$. Then, using Corollary 4.1.2, only the point $\lambda_0 = 3$ can be a bifurcation point. For that point, conditions **I** and **II** for $p = 1$ and $l = 3$ can be fulfilled. Then, following Theorem 4.1.4, a solution can be constructed as the following series:

$$x(t) = \sum_{n=1}^{\infty} a_n(t)\mu^{n/2},$$

where $\mu = \lambda - 3$. Using the method of undetermined coefficients, the following sequence of linear equations with respect to coefficients $a_n(t)$ can be obtained:

$$a_1(t) = ta_1(1),$$

$$a_2(t) = ta_2(1),$$

$$a_3(t) = ta_3(1) + ta_1(1) - \int_0^t tsa_1^3(s)\, ds,$$

$$a_4(t) = ta_4(1) + ta_2(1) - 3\int_0^t tsa_1^2(s)a_2(s)\, ds,$$

$$a_5(t) = ta_5(1) + ta_3(1) - 3\int_0^t tsa_1^2(s)a_3(s)\, ds,$$

$$\vdots$$

Then, in $a_1(t) = tc_1$, $a_2(t) = tc_2, \ldots$, the constant values c_1 and c_2 can be determined from the conditions of solvability for the subsequent equations of the chain. Indeed, c_1 can be determined from the algebraic equation $c_1 - \frac{c_1^3}{5} = 0$. Then, $c_1 = \pm\sqrt{5}$, and the integral equation has, in the positive semi-neighbourhood of the bifurcation point $\lambda_0 = 3$, two small nontrivial real solutions:

$$x_{1,2} \sim \pm t(\lambda - 3)^{1/2}\sqrt{5},$$

for $\lambda \to 3$.

Let us consider the following example:

$$x(t) = tb(\lambda)x(1) - \int_0^t (t-1)x^3(s)\, ds, \quad t \in [0,T], \quad T > 1.$$

Let $b(1) = 1$. Then, using Corollary 4.1.3, the point $\lambda_0 = 1$ is a bifurcation point. Moreover, for $\lambda_0 = 1$, a small solution for equation can be constructed as the series

$$x(t) = tc + \sum_{n=3}^{\infty} a_n(t)c^n.$$

This series converges uniformly for $t \in [0,T]$ and $|c| \le \rho$, where $\rho > 0$ is sufficiently small. The coefficients $a_n(t)$ can be calculated using the method of undetermined coefficients.

In order to estimate the convergence radius ρ, using the approach by Sidorov and Sidorov (2012), the following majorising system can be constructed:

$$\begin{cases} r = T\rho + T^2 r^3, \\ 1 = 3T^2 r^2. \end{cases}$$

The pair $r^* = \frac{1}{T\sqrt{3}}$ and $\rho^* = \frac{2}{3}\frac{r^*}{T}$, satisfies that system. Then, for $|c| \le \rho < \frac{2}{3\sqrt{3}T^2}$, the series converges uniformly, and the integral equation has a c-parametric solution in the space $\mathcal{C}_{[0,T]}$.

4.1.4 Volterra integral equations with vector load and vector bifurcation parameters

In equation (4.1), let x_α be a vector function: $x_\alpha = (x_{\alpha_1}, \ldots, x_{\alpha_n})$, $\lambda = (\lambda_1, \ldots, \lambda_m)$. Let us assume that both the load and the bifurcation parameter are in the neighbourhood of the zero of the vector spaces $\mathbb{R}^n$ and $\mathbb{R}^m$. Let us consider the equation $x(t) = \mathcal{L}(x(t), x_\alpha, t, \lambda)$, where

$$\mathcal{L}(x(t), x_\alpha, t, \lambda) = \sum_{k=1}^{n} b_k(t, \lambda_1, \ldots, \lambda_m)x_{\alpha_k}$$

$$+ \sum_{s=l}^{\infty} \sum_{j+k_1+\cdots+k_n=s} \left[f_{jk_1\ldots k_n}(t, \lambda_1, \ldots, \lambda_m)x^j(t)x_{\alpha_1}^{k_1}\ldots x_{\alpha_n}^{k_n} \right.$$

$$+ \left. \int_0^t K_{jk_1,\ldots,k_n}(t, s, \lambda_1, \ldots, \lambda_m)x(s)^j x_{\alpha_1}^{k_1}\ldots x_{\alpha_n}^{k_n}\, ds \right].$$

Here, b_k, $f_{jk_1,...,k_n}$, and $K_{jk_1,...,k_n}$ are continuous functions, which are sufficiently smooth with respect to λ_i in the neighbourhood $||\lambda - \lambda_0|| \leq \delta$.

Let us find the sufficient conditions for $\lambda_0 \in \mathbb{R}^m$ to be a bifurcation point. One can construct the asymptotics of real solutions for $\lambda_i = \lambda_{0i} + \mu$, $i = 1, \ldots, m$, where $\mu \in [0, \delta]$ (or for $\mu \in [-\delta, 0]$).

Let us start with the construction of the system for the vector load determination. To this end, let us construct a solution, $x(t)$, using the following series (4.10):

$$\sum_{j=1}^{\infty} a_j(t, \lambda, \boldsymbol{x}_\alpha). \tag{4.10}$$

Here,

$$a_j(t, \lambda, \mu x_{\alpha_1}, \ldots, \mu x_{\alpha_n}) = \mu^j a_j(t, \lambda, x_{\alpha_1}, \ldots, x_{\alpha_n}).$$

Note that j-homogeneous forms can be calculated uniquely using the following recurrence:

$$a_1(t, \lambda, x_\alpha) = \sum_{k=1}^{n} b_k(t, \lambda) x_{\alpha_k},$$

$$a_1 = \cdots = a_{l-1} = 0,$$

$$a_m(t, \lambda, x_\alpha) = \frac{1}{m!} \sum_{j+k_1+\cdots+k_n=m} \left[f_{jk_1,...,k_n}(t, \lambda) a_1(t, \lambda, x_\alpha)^j x_{\alpha_1}^{k_1} \ldots x_{\alpha_n}^{k_n} \right.$$

$$\left. + \int_0^t K_{jk_1,...,k_n}(t, s, \lambda) a_1(s, \lambda, x_\alpha)^j x_{\alpha_1}^{k_1} \ldots x_{\alpha_n}^{k_n} \, ds \right],$$

$m = l, l+1, \ldots, 2l-1;$

$$a_j(t, \lambda, x_\alpha) = \frac{1}{j!} \frac{\partial^j}{\partial \mu^j} \mathcal{L} \left(\sum_{n=1}^{j-1} a_n(t, \mu x_{\alpha_1}, \ldots, \mu x_{\alpha_n}), \mu x_\alpha, t, \lambda \right) \Bigg|_{\mu=0},$$

$j = 2l, 2l+1, \ldots.$

Remark 4.1.6. The sequence $\{x_m\}$, where $x_m(t) = \mathcal{L}(x_{m-1}(t), x_\alpha, t, \lambda)$ for the initial approximation $x_0(t) = \sum_{k=1}^{n} b_k(t, \lambda) x_{\alpha_k}$ and $||x_\alpha|| \leq \rho$, will uniformly converge to the sum of series (4.10). Then, in the nonanalytic case, the solution can be constructed using the successive approximation method.

Lemma 4.1.7. *The required load x_α in solution (4.10) must satisfy n nonlinear equations with parameters λ_i:*

$$A_1(\lambda)x_\alpha + \sum_{j=l}^{\infty} F_j(\lambda, x_\alpha) = 0. \tag{4.11}$$

Here, $F_j = (F_{j1}, \ldots, F_{jn})^T$, $\lambda = (\lambda_1, \ldots, \lambda_m)$, and F_{ji}, $i = 1, \ldots, n$ are the j-homogeneous forms of the components of the vector x_α, $A_1(\lambda) = [(b_k(t, \lambda), \alpha_i)]_{i,k=\overline{1,n}}$ is an $n \times n$ matrix, and α_i are linear Stieltjes functionals in the space $\mathcal{C}_{[0,T]}$.

The proof of the lemma is carried out by applying the functionals $\alpha_1, \ldots, \alpha_n$ to both sides of equality (4.10), taking into account the designations $x_\alpha = (x_{\alpha_1}, \ldots, x_{\alpha_n})$, where $x_{\alpha_i} = (x(t), \alpha_i)$, $i = 1, \ldots, n$, are functionals.

Considering system (4.11) leads to the following result.

Theorem 4.1.8 (necessary condition for bifurcation). *In order for the point $\lambda_0 \in \mathbb{R}^m$ to be a bifurcation point, it is necessary that $\det A_1(\lambda_0) = 0$.*

Proof. The point λ_0 can be a bifurcation point of the integral equation if and only if it is a bifurcation point of system (4.11). However, if $\det A_1(\lambda_0) \neq 0$, then system (4.11) has, besides the trivial solution, no other small solutions based on the implicit function theorem. Therefore, in view of formula (4.10), establishing a one-to-one correspondence between the small solution of the integral equation and a small solution of system (4.11), the integral equation will also have only a trivial small solution, $x(t) \equiv 0$, which completes the proof. $\qquad\square$

Sufficient condition for bifurcation in the vector case. Let us consider the application of the linearisation principle to derive the sufficient condition for bifurcation in the vector case.

Let the necessary bifurcation condition be satisfied (i.e. Theorem 4.1.8), and let $rank\, A_1(\lambda_0) = r$. Let $\{\phi_1, \ldots, \phi_{n-r}\}$ and $\{\psi_1, \ldots, \psi_{n-r}\}$ be orthonormal bases in the zero subspaces $\mathcal{N}(A_1(\lambda_0))$ and $\mathcal{N}(A_1^*(\lambda_0))$.

Let us introduce a matrix, $\widehat{A}_1(\lambda) = A_1(\lambda) + \sum_{i=1}^{n-r}(\cdot, \phi_i)\psi_i$, of dimension $n \times n$. Note that this square matrix is invertible because it is easy to show that the homogeneous system $\widehat{A}_1(\lambda_0)l = 0$ has only the trivial solution $l = 0$. Therefore, due to the continuity of $A_1(\lambda)$ with respect to λ, the matrix $\widehat{A}_1(\boldsymbol{\lambda})$ will also be invertible at least in the small neighbourhood $||\boldsymbol{\lambda} - \boldsymbol{\lambda}_0|| \leq \rho$ based on the inverse operator theorem.

Let us introduce the notation $\lambda_0 + \mu := (\lambda_{01} + \mu, \lambda_{02} + \mu, \ldots, \lambda_{0m} + \mu)$ and the auxiliary parameters $\xi_1, \ldots, \xi_{n-r}$ and rewrite system (4.11) in the form of the system of equations (4.12)–(4.13):

$$\widehat{A}_1(\lambda_0 + \mu)\boldsymbol{x}_\alpha = \sum_{k=1}^{n-r} \xi_k \psi_k - \sum_{j=l}^{\infty} F_j(\lambda_0 + \mu, \boldsymbol{x}_\alpha), \qquad (4.12)$$

$$\xi_i = (\boldsymbol{x}_\alpha, \phi_i), \quad i = 1, \ldots, n - r. \qquad (4.13)$$

Since $\boldsymbol{x}_\alpha \to 0$ for $\mu \to 0$, the parameters ξ_i will be infinitesimal. The matrix $\widehat{A}_1(\lambda_0 + \mu)$ has an inverse for $|\mu| < \rho$. Therefore, we can construct a convergent sequence, $\{\boldsymbol{x}_\alpha^n\}$, using recursion,

$$\boldsymbol{x}_\alpha^n = [\widehat{A}_1(\lambda_0 + \mu)]^{-1} \sum_{k=1}^{n-r} \xi_k \psi_k$$

$$- [\widehat{A}_1(\lambda_0 + \mu)]^{-1} \sum_{k=l}^{\infty} F_k(\lambda_0 + \mu, \boldsymbol{x}_\alpha^{n-1}),$$

upon the initial approach $\boldsymbol{x}_\alpha^0 = 0$. Then, $\lim_{n\to\infty} \boldsymbol{x}_\alpha^n = \boldsymbol{x}_\alpha$. Hence,

$$\boldsymbol{x}_\alpha = \sum_{k=1}^{n-r} [\widehat{A}_1(\lambda_0 + \mu)]^{-1} \psi_k \xi_k + R(\xi_1, \ldots, \xi_{n-r}, \mu), \qquad (4.14)$$

where

$$||R(\xi_1, \ldots, \xi_{n-r}, \mu)|| = \mathcal{O}(||\xi||^l).$$

Substituting the constructed $\boldsymbol{x}_\alpha$ into equalities (4.13), we obtain $n - r$ equations to determine the components of the vector $\xi(\mu)$:

$$\sum_{k=1}^{n-r} [(\widehat{A}_1(\lambda_0 + \mu)^{-1}\psi_k, \phi_i) - \delta_{ki}]\xi_k + r_i(\xi, \mu) = 0, \qquad (4.15)$$

$i = 1, \ldots, n - r,$

$$|r_i(\xi, \mu) = \mathcal{O}(||\xi||^l)|.$$

Let us find the sufficient conditions independent of the form of nonlinear functions $r_i(\xi, \mu)$, under which $\mu = 0$ will be the bifurcation point of system (4.15).

Let us introduce a continuous function:

$$\Delta(\mu) = \det[(\widehat{A}_1(\lambda_0 + \mu)^{-1}\psi_k, \phi_i) - \delta_{ki}]_{i,k=1}^{n-r}. \tag{4.16}$$

In view of the identity $\widehat{A}_1(\lambda_0)\phi_k = \psi_k$, we have the following identity: $\widehat{A}_1^{-1}(\lambda_0)\psi_k = \phi_k$, $k = 1, \ldots, n - r$. Therefore, $\lim_{\mu \to 0} \Delta(\mu) = 0$, and the point $\mu = 0$ is the zero of the function $\Delta(\mu)$. Let us note the following condition:

V. Let $\mu = 0$ be the k-multiple zero of the function $\Delta(\mu)$.
Then, $\Delta(\mu) \sim c\mu^k$ for $\mu \to 0$ and $\frac{d^i}{d\mu^i}\Delta(\mu)|_{\mu=0} = 0$, $i = 0, \ldots, k - 1$, $\frac{d^k \Delta(\mu)}{d\mu^k}\big|_{\mu=0} \neq 0$. If the matrix $A_1(\lambda+\mu)$ at an isolated singular point λ_0 has a complete generalised Jordan set (see item 30 in Vainberg and Trenogin, 1964) of $n - r$ chains of length p_i, $i = 1, \ldots, n - r$, then $k = p_1 + \cdots + p_{n-r}$.

Definition 4.1.2. Let us call the number k the root number of the matrix $A_1(\lambda)$ at point λ_0.

Chapter 9 of the monograph by Vainberg and Trenogin (1964) shows that the root number is independent of the choice of bases in $\mathcal{N}(A(\lambda_0))$ and $\mathcal{N}(A^*(\lambda_0))$ at the Fredholm singular point λ_0.

Theorem 4.1.9 (linearisation principle). *Let the root number k of the matrix $A_1(\lambda)$ at the point λ_0 be odd. Then, λ_0 will be the bifurcation point of the integral equation (4.1) with vector load.*

Proof. In view of formulas (4.10) and (4.14), λ_0 will be the bifurcation point of equation (4.1) in the vector case if and only if $\mu = 0$ is a bifurcation point of system (4.15). We write system (4.15) in vector form as

$$S(\mu)\xi + r(\xi, \mu) = 0, \tag{4.17}$$

where $S(\mu) = [(\widehat{A}_1(\lambda_0 + \mu)^{-1}\psi_k, \phi_i) - \delta_{ki}]_{k,i=1,\ldots,n-r}$ is a square matrix, $\xi \in \mathbb{R}^{n-r}$, $\|r(\xi, \mu)\| = \mathcal{O}(\|\xi\|^l)$. Let us set an arbitrarily small $\delta > 0$ and $\varepsilon > 0$ and introduce the vector field

$$H(\xi, t) := S((2t - 1)\delta)\xi + r(\xi, (2t - 1)\delta) : \mathbb{R}^{n-r} \times \mathbb{R}^1 \to \mathbb{R}^{n-r},$$

defined for $\xi, t \in M$, where $M = \{\xi, t \mid \|\xi\| < \varepsilon, 0 \le t \le 1\}$.

Case 1.
If there is a pair $(\xi^*, t^*) \in M$, for which $H(\xi^*, t^*) = 0$, then, according to Definition 4.1.1, the point $\mu = 0$ will be the bifurcation point of system (4.15), and Theorem 4.1.9 is true.

Case 2.

Suppose that $H(\xi, t) \neq 0$ for any $(\xi, t) \in M$; therefore, Theorem 4.1.9 is incorrect. Due to the continuity of the field $H(\xi, t)$ for $\xi, t \in M$, $H(\xi, 0)$ and $H(\xi, 1)$ are homotopic on the sphere $||\xi|| = \varepsilon$; therefore, their rotations on this sphere are equal, i.e.

$$J(H(\xi, 0), ||\xi|| = \varepsilon) = J(H(\xi, 1), ||\xi|| = \varepsilon). \tag{4.18}$$

Due to the smallness, ε, of the fields $H(\xi, 0)$ and $H(\xi, 1)$ homotopic to their non-degenerate linear parts $S(-\delta)\xi$ and $S(+\delta)\xi$,

$$J(H(\xi, 0), ||\xi|| = \varepsilon) = J(S(-\delta)\xi, ||\xi|| = \varepsilon), \tag{4.19}$$

$$J(H(\xi, 1), ||\xi|| = \varepsilon) = J(S(+\delta)\xi, ||\xi|| = \varepsilon). \tag{4.20}$$

Due to condition **V**, $\det S(-\delta) \sim c(-\delta)^k$ and $\det S(+\delta) \sim c(+\delta)^k$. Therefore, by the Kronecker index theorem,

$$J(S(-\delta)\xi, ||\xi|| = \varepsilon) = \operatorname{sign}(-\delta)^k,$$

$$J(S(+\delta)\xi, ||\xi|| = \varepsilon) = \operatorname{sign}(+\delta)^k.$$

Since k is odd, we arrive at the equalities

$$J(S(-\delta)\xi, ||\xi|| = \varepsilon) = -1,$$

$$J(S(+\delta)\xi, ||\xi|| = \varepsilon) = +1,$$

and equality (4.18) becomes impossible! Therefore, for odd k, there is always a pair, $(\xi^*, t^*) \in M$, for which $H(\xi^*, t^*) = 0$, and according to Definition 4.1.1, the point $\mu = 0$ will be the bifurcation point of system (4.15), and the point λ_0 will be, accordingly, the bifurcation point of the integral equation (4.1) in the vector case. $\qquad \square$

Remark 4.1.10. From Theorem 4.1.9, in the case of a single load, the validity of Theorem 4.1.5 follows as a consequence.

Corollary 4.1.11. *In equation* (4.1), *let the load* x_α *and parameter* λ *be vectors,* $\det A_1(\lambda_0) = 0$, *rank* $A_1(\lambda_0) = r$, $\{\phi_i\}_1^{n-r}$, *and* $\{\psi_i\}_1^{n-r}$ *be bases in the subspaces* $\mathcal{N}(A_1(\lambda_0))$ *and* $\mathcal{N}(A_1^*(\lambda_0))$. *If* $(n - r)$ *is an odd number and* $\det[(A_1^{(1)}(\lambda_0)\phi_1, \psi_k)]_{i,k=1}^{n-r} \neq 0$, *then* λ_0 *is the bifurcation point of equation* (4.1).

Proof. Under the conditions of Corollary 4.1.11, at an isolated Fredholm point λ_0, the matrix $A_1(\lambda_0 + \mu)$ for $|\mu| < r$ has no elements attached. Therefore, the vectors $\phi_1, \ldots, \phi_{n-r}$ constitute its complete generalised Jordan set, and the root number k is equal to $n - r$. Thus, the conditions of Theorem 4.1.9 are satisfied. □

Corollary 4.1.12. *Let equation* (4.1) *have a load* x_α *and a parameter* λ *as vectors,* $\det A_1(\lambda_0) = 0$, $\operatorname{rank} A_1(\lambda_0) = r$ $\{\phi_i\}_1^{n-r}$, *and* $\{\psi_i\}_1^{n-r}$ *be bases in the subspaces* $\mathcal{N}(A_1(\lambda_0))$ *and* $\mathcal{N}(A_1^*(\lambda_0))$. *Let*

$$\frac{d^j}{d\lambda^j} A_1(\lambda)\phi_i \Big|_{\lambda=\lambda_0} = 0$$

for $j = 0, \ldots, p-1$, $i = 1, \ldots, n-r$,

$$\det \left[\frac{d^p}{d\lambda^p} \langle A_1(\lambda)\phi_i, \psi_j \rangle \right]_{i,j=1}^{n-r} \Bigg|_{\lambda=\lambda_0} \neq 0,$$

and the number $p(n-r)$ *is odd. Then,* λ_0 *is the bifurcation point of equation* (4.1).

Proof. Under the conditions of Corollary 4.1.12, the function $\Delta(\mu)$, defined above in formula (4.16), point $\mu = 0$ is $p(n - r)$-divisible by zero. Corollary 4.1.12 follows from the proof of the linearisation principle because $p(n - r)$ is odd number.

The linearisation principle is a sufficient condition for the existence of a bifurcation point, the verification of which uses only information about the linear parts of equation (4.1). When the conditions of Theorem 4.1.9 are met, equation (4.1) with n loads can have a variety of small real solutions $x(t, \lambda_0)$, depending on $n - r$ small parameters, where r is the rank of a certain matrix and λ_0 is a bifurcation point.

Further, as above, we obtain in the vector case sufficient conditions when nontrivial real solutions, $x(t, \lambda) \to 0$ for $\lambda \to \lambda_0$, can be constructed in at least one of the semi-neighbourhoods of the bifurcation point using additional information about the nonlinear part of equation (4.1).

In view of formula (4.10), which states a one-to-one correspondence between the solutions to equation (4.1) and the solutions to system (4.11), to solve this problem, it is enough to be able to construct a real solution, $x_\alpha \to 0$, at $\lambda \to \lambda_0 + 0$ ($\lambda \to \lambda_0 - 0$) for system (4.11).

Let us introduce conditions on system (4.11). For $\mu \to 0$, let the following asymptotic estimates be valid for $\mu \to 0$:

VI. $A_1(\lambda_0 + \mu)\boldsymbol{x}_\alpha \sim F_{1p_1}(\lambda_0)^{-1}\boldsymbol{x}_\alpha \mu^{p_1}$,

$F_i(\lambda_0 + \mu, \boldsymbol{x}_\alpha) \sim F_{ip_i}(\lambda_0, \boldsymbol{x}_\alpha)\mu^{p_i}, i = l, l+1, \ldots$.

Moreover, $p_i > 0$, $i = 1, l, l+1, \ldots, n-1, p_n = 0$, F_{1p_1} is an $n \times n$ matrix, and $F_{ip_i}(\lambda_0, \boldsymbol{x}_\alpha)$ is an i-homogeneous forms with respect to the vector components load $\boldsymbol{x}_\alpha$.

As in the case of one load, by plotting the points $(1, p_1)$, (l, p_l), $(l+1, p_{l+1}), \ldots, (n-1, p_{n-1})$, $(n, 0)$ on the coordinate plane, we build a Newton's diagram, and we look for a load vector of the form

$$\boldsymbol{x}_\alpha = \boldsymbol{c}(\mu)|\mu|^{r/s}, \quad \boldsymbol{c}(0) \neq 0.$$

To determine the vector $\boldsymbol{c}(0)$, we obtain two different systems, one for the case $\mu > 0$ and the other for $\mu < 0$:

$$P_\pm(\boldsymbol{c}) := \sum_{i\frac{r}{s}+k=\theta} \operatorname{sign}\mu^k F_{ik}(\lambda_0, \boldsymbol{c}) = 0. \tag{4.21}$$

The components of the vectors $F_{ik}(\lambda_0, \boldsymbol{c})$ are i-homogeneous forms with respect to the vector $\boldsymbol{c}$. By constructing a simple non-zero real solution, $\boldsymbol{c}$, of one of these systems, it will be possible in the corresponding semi-neighbourhood of the bifurcation point λ_0 to write down the asymptotics $x(t, \lambda) \sim \sum_{k=1}^n b_k(t, \lambda_0)c_k^*|\lambda - \lambda_0|_s^r$ for $\lambda \to \lambda_0 + 0$ ($\lambda \to \lambda_0 - 0$) of solutions to the integral equation (4.1). □

The presented method allows us to obtain, for $\lambda \in \mathbb{R}^m$, the sufficient conditions for the existence of more general varieties, namely curves and bifurcation surfaces for the integro-differential equation

$$\begin{cases} \dfrac{dx}{dt} + a(t, \lambda)x(t) = \mathcal{L}(x(t), x_\alpha, t, \lambda), \\ x(0) = 0, \end{cases}$$

with loads. The method allows one to analyse more general classes of differential operator equations in Banach spaces with loads in the vicinity of branch points of the solutions.

The bifurcation analysis of the solution to the Cauchy problem for a differential equation with loads is introduced. As a aside, it should be noted that the results of this chapter can be applied in the analysis of non-standard nonlinear integral models with loads arising in the study of heriditarian processes.

4.2 Branching Solutions of the Cauchy Problem for Nonlinear Loaded Differential Equations with Bifurcation Parameters

The development of advanced methods and models for nonlinear dynamical systems control depends on the theory of loaded systems of equations (Abdullaev and Aida-Zade, 2014; Aidazade and Abdullaev, 2014; Dikinov *et al.*, 1976; Nahushev, 2012). Indeed, there is a great variety in the types of loads that must be taken into account both in the design of technical systems and in the creation of adequate models for complex energy systems. Loaded ordinary differential equations model heat transfer phenomena and are solved using the finite difference method by Alikhanov *et al.* (2008).

It is worth noting that loaded differential and loaded integro-differential equations are directly relevant to the non-local problem for integro-differential equations; see Baltaeva (2017), Agarwal *et al.* (2020), and the references therein. Nevertheless, the problem of analysing loaded systems of differential equations with bifurcation parameters remains an open problem despite the significant progress in nonlinear analysis (Sidorov, 2022).

The purpose of this section is to report the general existence theorems and construct approximate methods for solving the nonlinear loaded Cauchy problem with bifurcation parameters.

First, a brief introduction and a problem statement are given. The Cauchy problem for a single equation is considered in Sections 4.2.1 and 4.2.2. Necessary and sufficient conditions are given, under which the point λ^0 will be a bifurcation point, and a method for constructing a real solution is proposed. The theoretical results are illustrated by solving a differential equation. In Section 4.2.3, two theorems on the existence of bifurcation points in the Cauchy problem for systems of equations are formulated.

Let us consider the following Cauchy problem:

$$\begin{cases} \dfrac{dx(t)}{dt} = \Phi(x(t), x_\alpha, t, \lambda), \\ x(0) = 0. \end{cases} \tag{4.22}$$

The desired vector function $x(t) := (x_1(t), \ldots, x_n(t))^T$ is continuous for $t \in [0, T]$ and the parameter $\lambda \in \Omega \subset E^m$, where E^m is a normed space. The load $x_\alpha := (x_{\alpha_1}, \ldots, x_{\alpha_n})^T$ is given using the linear Riemann–Stieltjes functionals

$$x_{\alpha_i} := \int_{a_1}^{b_i} x_i(t) \, d\alpha_i(t), \quad i = 1, \ldots, n,$$

where $\alpha_i(t)$ is a bounded variation function, $[a_i, b_i] \subset (0, T)$. Such functionals, x_α, are called *integral loads*. Likewise, the functional $x_\alpha := (x(\alpha_1), \ldots, x(\alpha_n))^T$, where $a_i \in (0, T)$, is called a *local load*.

The function Φ in system (4.22) is continuous and defined as $\Phi(x(t), x_\alpha, t, \lambda) = (\Phi_1(x(t), x_\alpha, t, \lambda), \ldots, \Phi_n(x(t), x_\alpha, t, \lambda))^T$, with $\Phi_i = \sum_{j=1}^n a_{ij}(t, \lambda) x_{\alpha_j} + \sum_{j=l}^N F_{ij}(x(t), t) + o(||x||^N)$, where $i = 1, \ldots, n$, $l \geq 2$, $F_{ij}(x(t), t)$ are j-homogeneous (normal) forms of the vector function $x(t)$. System (4.22) for an arbitrary parameter λ has the trivial solution $x(t) = 0$, $x_\alpha = 0$.

The goal is to find the points λ^0 such that the Cauchy problem (4.22) has a nontrivial real solution in the neighbourhood of λ^0. Such values of the parameter are usually called *bifurcation points* (branching points) of solutions. In applications, the parameter λ may be associated with an external influence imposed on the system. Therefore, bifurcation points are of particular interest in mathematical modelling. Note that loaded equations model dynamic processes with trajectories at any fixed points in time (or local time intervals) influencing the whole trajectory; however, there is no practical (technical) way to perform measurements at these points or intervals.

There have been a number of studies on loaded differential equations (Aidazade and Abdullaev, 2014; Dikinov *et al.*, 1976; Nahushev, 2012). In particular, the nonlinear Volterra integral equation with local and/or integral loads on the desired solution given by the Stieltjes integral was considered by Sidorov and Sidorov (2021). The Cauchy problem (4.22) contains a bifurcation parameter, λ, and has a trivial solution for any of the parameter values. Necessary and sufficient conditions are obtained for those values of the parameter (bifurcation points) in the neighbourhood of which problem (4.22) has nontrivial continuous real solutions. In this section, we summarise these studies since bifurcation analyses of Cauchy problems represented by systems of differential equations with loads have not been conducted yet.

Definition 4.2.1. The arbitrary bounded domain $S \subset E^m$ is called the sectorial neighbourhood of the point λ^0 if $\lambda^0 \in \partial S$.

Definition 4.2.2. The point λ^0 is called the *bifurcation point* of the initial problem (4.22) if, for arbitrary $\varepsilon > 0$ and $\delta > 0$, there exist $x(t)$ and λ in the sectorial neighbourhood of point λ^0 satisfying problem (4.22) such that $0 < ||x|| \leq \varepsilon$, $||\lambda - \lambda^0|| \leq \delta$.

In this section, the necessary and sufficient conditions for problem (4.22) to possess a bifurcation point are derived. In this case, one can construct asymptotics of non-trivial real branches of small solutions for system (4.22) in the vectorial domain of this point. By a *branch*, we mean a continuous real solution $x(t)$ such that $x(t) \to 0$ as $S \ni \lambda \to \lambda^0$.

Clearly, each solution $x(t)$ of problem (4.22) is assigned a single load using a given linear functional. We obtain a solution to the problem by constructing an equation with respect to the load with the parameter λ and investigate it using the Lyapunov and Schmidt ascending approach based on a combination of methods:

(1) the analytical method of successive approximations and the Nekrasov–Nazarov method of uncertain coefficients;
(2) the geometric method based on the Kronecker–Poincaré index and the power geometry of Newton's diagrams (polygons);
(3) the homotopy perturbation method.

The solution methodology is constructive, does not employ any complex generalisations, and is accessible to a wide range of specialists in applied fields.

4.2.1 *Construction of equations with respect to load: The necessary conditions for bifurcation and the existence of solutions*

For the sake of clarity, let us start with the simple case of a single equation:

$$\begin{cases} \dfrac{dx}{dt} = a(t, \lambda) x_\alpha + \displaystyle\sum_{n=l}^{\infty} F_n(t) x^n(t), \\ x(0) = 0, \end{cases} \qquad (4.23)$$

with a single load x_α and parameter $\lambda \in \mathbb{R}^1$. As we are interested in small solutions, the desired solution $x(t)$ as a function of the load can be constructed as the following series:

$$x(t) = a_1(t, \lambda) x_\alpha + \sum_{i=l}^{\infty} a_i(t, \lambda) x_\alpha^i. \qquad (4.24)$$

In a non-analytic case, the method of successive approximations can be applied. The series coefficients can be calculated using the method of undetermined coefficients by the substitution of series (4.24) into the equation

$$x(t) = \int_0^t a(t, \lambda)\, dt x_\alpha + \sum_{n=l}^\infty \int_0^t F_n(t) x^n(t)\, dt.$$

Then, clearly, we find the recursive formulas

$$a_1(t, \lambda) = \int_0^t a(t, \lambda)\, dt,$$

$$a_l(t, \lambda) = \int_0^t F_l(t) a_1^l(t, \lambda)\, dt,$$

$$\vdots,$$

and we can construct the uniformly converging series (4.24) for small enough $|x_\alpha|$. Using the standard notion of $\langle x, \alpha \rangle$ and by the application of the functional to both sides of formula (4.24), the following equation with respect to load x_α can be derived:

$$L(x_\alpha, \lambda) := L_1(\lambda) x_\alpha + \sum_{i=l}^\infty L_i(\lambda) x_\alpha^i = 0, \qquad (4.25)$$

where $L_1(\lambda) = \langle a_1(t, \lambda), \alpha \rangle - 1$, $L_i(\lambda) = \langle a_i(t, \lambda), \alpha \rangle$, $i = l, l+1, \ldots$. Equation (4.25) is called a *branching equation* with respect to the load.

Thus, the following statement is true.

Lemma 4.2.1. *Load x_α in problem (4.23) must satisfy equation (4.25) with parameter λ. The bifurcation points of the Cauchy problem (4.23) will be the bifurcation points of the branching equation (4.25).*

Then, from Lemma 4.2.1 and the implicit function theorem, there follow two corollaries.

Corollary 4.2.2 (necessary conditions for a bifurcation). *In order for point λ^0 to be the bifurcation point of problem (4.23), it is necessary that $L_1(\lambda^0) = 0$ in equation (4.25).*

Corollary 4.2.3. *In equation (4.25), let all the coefficients $L_i(\lambda)$, $i = 1, l, l+1, \ldots$, in point λ^0 be zeros. Then, λ^0 is the bifurcation point of equation (4.23). Moreover, equation (4.23), for $\lambda = \lambda^0$, will enjoy the c-parametric nontrivial solution $x(t, c) = a_1(t, \lambda^0)c + \sum_{i=l}^\infty a_i(t, \lambda^0)c^i$, depending on a sufficiently small parameter c. For $0 < |\lambda - \lambda^0| < \delta$, where δ is sufficiently small, there are no other small solutions for problem (4.23).*

4.2.2 Sufficient conditions for the existence of bifurcation points and solution asymptotics

By Lemma 4.2.1, to construct solutions for equation (4.23), in equation (4.25), we must use $\lambda = \lambda^0 + \mu$, where μ is a small real parameter, and construct the asymptotics of the small solution $x_\alpha \to 0$ as $\mu \to 0$ for equation (4.25). For this purpose, we introduce the following condition:

I. λ^0 is an odd root of equation $L_1(\lambda) = 0$ of multiplicity p, with $L_l(\lambda^0) \neq 0$.

Then, for $\lambda = \lambda^0 + \mu$, equation (4.25) can be transformed as follows:
$\left(\frac{1}{p!}L_1^{(p)}(\lambda^0)\mu^p + \mathcal{O}(|\mu|^{p+1})\right)x_\alpha + \left(L_l(\lambda^0) + \mathcal{O}(|\mu|)x_\alpha^l + \mathcal{O}(|x_\alpha|^{l+1})\right) = 0$ in the neighbourhood of the points $x_\alpha = 0$ and $\mu = 0$. By applying a Newton's diagram (Vainberg and Trenogin, 1964), we can construct the load as $x_\alpha = (c_0 + \mathcal{O}(|\mu|))\mu^{\frac{p}{l-1}}$. If p is odd, then c_0 satisfies, for $\mu > 0$, the following equation:

$$\frac{1}{p!}L_1^{(p)}(\lambda^0) + L_l(\lambda^0)c_0^{l-1} = 0,$$

and for $\mu < 0$, it satisfies

$$-\frac{1}{p!}L_1^{(p)}(\lambda^0) + L_l(\lambda^0)c_0^{l-1} = 0.$$

One of these equations for odd l has a simple nontrivial real solution. Then, the real c_0 can always be constructed explicitly, and one can construct the main term of the asymptotics of the solution of the Cauchy problem (4.23).

Therefore, we get the following theorem.

Theorem 4.2.4. *Let condition **I** be satisfied. Then, λ^0 is the bifurcation point of equation (4.23). Moreover, for even l and arbitrary p, equation (4.23) enjoys a unique small real solution $x(t, \lambda)$ in the neighbourhood of point λ^0, and*

$$x(t, \lambda) \sim \int_0^t a(t, \lambda^0)\, dt \sqrt[l-1]{-\frac{1}{p!}\frac{L_1^{(p)}(\lambda^0)}{L_l(\lambda^0)}(\lambda - \lambda^0)^p}.$$

For odd l and odd p, equation (4.23) has a real solution with a similar asymptotic in one-half neighbourhood of point λ^0.

Let us consider the following problem:

$$\begin{cases} \dfrac{dx(t)}{dt} = (1 - \lambda)x(\alpha) + bx^3(t), \quad t \in \mathbb{R}^1, \\ x(0) = 0. \end{cases} \qquad (4.26)$$

The real value α is given. This example satisfies the conditions of Theorem 4.2.4 for $p = 1$ and $l = 3$.

For sufficiently small $|x(\alpha)|$, its solution, according to decomposition (4.24), is represented as a function of the load $x(\alpha)$ in the form of a series: $x(t) = t(1 - \lambda)x(\alpha) + \frac{b}{4}t^4(1 - \lambda)^3 x(\alpha)^3 + \mathcal{O}(|x(\alpha)|^4)$. Then, the branching equation (4.25) in this example is as follows:

$$[\alpha(1 - \lambda) - 1]x(\alpha) + \frac{b}{4}\alpha^4(1 - \lambda)^3 x(\alpha)^3 + \mathcal{O}(|x(\alpha)|^4) = 0.$$

The load x_α is defined from equation (4.25), with $L_1(\lambda) = \alpha(1 - \lambda) - 1$ and $L_3(\lambda) = \frac{b}{4}\alpha^4(1 - \lambda)^3$. Then, due to Corollary 4.2.2, $\lambda^0 = 1 - \frac{1}{\alpha}$ is the unique bifurcation point.

Using the Newton's diagram method, the branching equation (4.25) has the solution in the form of a power series with powers of $\mu^{1/2}$. Therefore, using formula (4.24), the two branches of the nontrivial real solution of the Cauchy problem should be sought in the form of a series:

$$x(t) = \sum_{n=1}^{\infty} c_n(t)\mu^{n/2}, \quad c_n(0) = 0, \quad n = 1, 2, \dots. \qquad (4.27)$$

To determine the coefficients $c_n(t)$, we obtain the following recurrent sequence of initial problems:

$$\begin{cases} \dfrac{dc_1(t)}{dt} = \dfrac{1}{\alpha}c_1(\alpha), \\ c_1(0) = 0, \end{cases}$$

$$\begin{cases} \dfrac{dc_2(t)}{dt} = \dfrac{1}{\alpha}c_2(\alpha), \\ c_2(0) = 0, \end{cases}$$

$$\begin{cases} \dfrac{dc_3(t)}{dt} = \dfrac{1}{\alpha}c_3(\alpha) - c_1(\alpha) + bc_1^3(t), \\ c_3(0) = 0, \end{cases}$$

$$\vdots$$

Then, $c_1(t) = \frac{1}{\alpha}c_1(\alpha)t$ and $c_2(t) = \frac{1}{\alpha}c_2(\alpha)t$. The constant value $c_1(\alpha)$ can be uniquely defined from a third-level solvability condition. Indeed, the third equation is given by

$$\begin{cases} \dfrac{dc_3(t)}{dt} = \dfrac{1}{\alpha}c_3(\alpha) - c_1(\alpha) + \dfrac{b}{\alpha^3}c_1^3(t)t^3, \\ c_3(0) = 0. \end{cases}$$

Then, $c_3(t) = \frac{t}{\alpha}c_3(\alpha) - c_1(\alpha)t + \frac{bc_1(\alpha)^3}{\alpha^3}\frac{t^4}{4}$. The solvability condition is as follows: $-c_1(\alpha)\alpha + bc_1(\alpha)^3\frac{\alpha}{4} = 0$. Then, $c_1(\alpha) = \pm\frac{2}{\sqrt{b}}$, and the Cauchy problem (4.26) due to (4.27) enjoys two real solutions in the semi-neighbourhood of the point $\lambda^0 = 1 - \frac{1}{\alpha}$, with asymptotics

$$x_{1,2}(t) \sim \pm\frac{2t}{\alpha}\sqrt{\frac{\lambda - \lambda^0}{b}}.$$

The sign of the semicircle in which the solution is real coincides with the sign of the coefficient b, i.e. for $(\lambda - \lambda^0)/b > 0$.

Asymptotics can be specified by determining the coefficients of series (4.27). To determine the arbitrary constants appearing in the solution of the nth linear boundary value problem, the solvability conditions of $n + 2$ equations of the recurrence chain of equations are used. Thus, as in the Nekrasov–Nazarov method (Vainberg and Trenogin, 1964; Akhmedov, 1957) developed for problems in mechanics, in our case, the calculation methodology has two steps and uses the Newton–Puiseux expansion. This method, considering the recent results of power geometry (Bruno, 2000) and the methods of nonlinear analysis, can also be applied to the study of the Cauchy problem with load vector and vector bifurcation parameters.

In Section 4.2.3, we present a general theorem for the existence of bifurcation points of system (4.22) in the vector case. Theorem 4.2.8 is a justification for possible linearisation in the obtained sufficient condition for the existence of bifurcation points. For the theory of nonlinear equations with vector parameters, readers may refer to the book by Leontyev (2013). Theorem 4.2.8 provides a method for checking for bifurcation using only the linearisation of the Cauchy problem (4.22). A similar result for other nonlinear problems was first proven by Krasnoselsky (1964).

4.2.3 *Necessary and sufficient conditions for the existence of a nontrivial solution of the Cauchy problem for systems in the vector case*

As in the case of a single equation, the desired solution of system (4.22) will be plotted as a function of the load vector. In this section,

$x(t) = (x_1(t), x_2(t), \ldots, x_n(t))^T$, $\lambda = (\lambda_1, \lambda_2, \ldots, \lambda_m)^T$, and $x_\alpha = (x_{\alpha_1}, x_{\alpha_2}, \ldots, x_{\alpha_n})^T$.

Let us build the sequence $x_n = \int_0^t \Phi(x_{n-1}(t), \lambda, x_\alpha) \, dt$, $n = 1, 2, \ldots$, at the initial approach

$$x_0(t) = \sum_{j=1}^n \int_0^t a_{ij}(s, \lambda) \, ds x_{\alpha_j} \Big|_{j=1}^n .$$

For a sufficiently small norm $||x_\alpha||$, this sequence converges, and its limit represents the solution of the Cauchy problem (4.22) expressed in terms of the load vector x_α. The load vector x_α can be found from a system of small implicit functions:

$$L(x_\alpha, \lambda) := L_1(\lambda) x_\alpha + \mathcal{O}(||x_\alpha||^l) = 0, \qquad (4.28)$$

where

$$L_1(\lambda) = \left[\int_{a_i}^{b_i} \int_0^t a_{ij}(s, \lambda) \, ds \, d\alpha_i(t) - \delta_{ij} \right]_{i,j=1}^n$$

is a square matrix.

For the convenience of calculations, we use the notation

$$L_1(\lambda) := \left[A_{ij}(\lambda) \right]_{i,j=1}^n .$$

The bifurcation points of system (4.28) will be the desired bifurcation points of the Cauchy problem (4.22).

Let us introduce $D = \{\lambda | \det L_1(\lambda) = 0\} \subset \Omega$. Based on the implicit mapping theorem, the following Lemma is valid.

Lemma 4.2.5 (necessary bifurcation conditions for the Cauchy problem). *For the point λ to be a bifurcation point of the Cauchy problem (4.22), it is necessary to fulfill the inclusion $\lambda^0 \in D$.*

To obtain sufficient bifurcation conditions in problem (4.22), we need the following condition:

(A) Let $\lambda^0 \in D$, and there exists its sectorial neighbourhood $S = S_+ \cup S_-$ and $\lambda^0 \in \partial S_+ \cap \partial S_-$.

Let l be a fixed vector from the set Ω, and there exists a continuous mapping $\lambda(\mu) := \lambda^0 + l\mu$, with $\mu = (2\theta - 1)\delta$, where δ is a small enough value, $\delta > 0$, $\theta \in [0, 1]$,

$$\det[L_1(\lambda)] = \alpha(\mu),$$

where $\alpha(\mu)$ is a continuous function, $\alpha(\mu) < 0$ for $\mu \in [-\delta, 0)$, and $\alpha(\mu) > 0$ for $\mu \in (0, +\delta]$, $\alpha(0) = 0$.

It is to be noted that μ is a small parameter here. The following lemma formulates the sufficient conditions, thereby fulfilling condition **(A)**.

Lemma 4.2.6. *Let λ^0 and l be fixed vectors from the normed space E^m. μ is a small real parameter,*

$$A_{ik}(\lambda^0 + \mu l) = b_{ik}\mu^{p_k} + o(|\mu|^{p_k})$$

*for $i, k = 1, \ldots, n$, and $p_k \in \mathbb{N}$, $\det[b_{ik}]_{i,k=1}^n \neq 0$, $p_1 + \cdots + p_n$ are odd numbers. Then, condition **(A)** is fulfilled, and*

$$\alpha(\mu) \sim \mu^{p_1 + \cdots + p_n} \det[b_{ik}]_{i,k=1}^n$$

for $\mu \to 0$.

Proof. Since $p_k > 0$, $k = 1, \ldots, n$, $\lambda^0 \in D$, and under Lemma 4.2.6, the following estimate is valid:

$$||L_1(\lambda^0 + \mu l) - [b_{ik}\mu^{p_k}]_{i,k=1}^n|| = o(|\mu|^{p_1 + \cdots + p_n}).$$

The matrix $[b_{ik}\mu^{p_k}]_{i,k=1}^n$ is the product of the non-degenerate matrix $[b_{ik}]_{i,k=1}^n$ and the diagonal matrix $[\delta_{ik}\mu^{p_k}]_{i,k=1}^n$. Since the determinant of the product of the two matrices is the product of the determinants of these matrices, due to the above-outlined estimate, the following estimate is true:

$$\det[A_{ij}(\mu)]_{i,j=1}^n \sim \mu^{p_1 + \cdots + p_n} \det[b_{ik}]_{i,k=1}^n.$$

The lemma is proven. □

Remark 4.2.7. Lemma 4.2.6 strengthens the results presented in the paper by Sidorov and Trenogin (2003) for the case of equations with loads.

Using the above lemma and the methods in the paper by Sidorov and Trenogin (2003), we obtain the following result.

Theorem 4.2.8. *Let condition **(A)** be satisfied. Then, λ^0 is the bifurcation point of problem (4.22).*

Proof. Let us define arbitrary small $\varepsilon > 0$ and $\delta > 0$ and introduce a continuous vector field,

$$H(x_\alpha, \theta) := L(x_\alpha, \lambda^0 + l(2\theta - 1)\delta) : \mathbb{R}^n \times \mathbb{R}^1 \to \mathbb{R}^n,$$

defined for $x_\alpha, \theta \in M$, where

$$M = \{x_\alpha, \theta \mid ||x_\alpha|| = \varepsilon, \, 0 \le \theta \le 1\}.$$

Let us assume that the theorem is not valid, i.e. $H(x_\alpha, \theta) \ne 0$ for arbitrary x_α, θ from set M. Then, the fields $H(x_\alpha, 0)$ and $H(x_\alpha, 1)$ are homotopic on the sphere $||x_\alpha|| = \varepsilon$. Moreover, they are homotopic for sufficiently small ε with respect to their linearisation and non-homogeneous case. However, that is impossible due to condition **(A)**, the inequality $\alpha(\mu) < 0$ for $\mu \in [-\delta, 0)$ and $\alpha(\mu) > 0$ for $\mu \in (0, +\delta]$. Then, in set M, there exists a pair (x_α^*, θ^*) such that $H(x_\alpha^*, \theta^*) = 0$, and λ^0 is a bifurcation point. $\square$

Many works are devoted to the study of bifurcation points of non-linear equations. From a computational point of view, the bifurcation point problem is uncorrected in the Tikhonov–Lavrentiev sense. A small perturbation in the original equation can lead to significant changes in the structure of the solution. In particular, the perturbed problem may have no real solutions at all in a sufficiently small neighbourhood of the bifurcation point. Note that the presence of a bifurcation parameter in the equation significantly complicates the problem of constructing stable computations since we should consider 'parameter-equal regularisation'. The monographs Sidorov *et al.* (2002). Sidorov *et al.* (2020c) provide a systematic account of a number of sections and applications of the modern theory of branching nonlinear equations. The presentation is based on the reduction of the original nonlinear problem to an equivalent finite-dimensional problem. In this section, this approach, going back to Lyapunov and Schmidt, using a combination of some recent results of nonlinear analysis, is effectively applied to constructing solutions for classes of nonlinear differential equations with stresses. In the considered Cauchy problem, the role of such a finite-dimensional problem is played by the system with respect to loads.

The studies conducted by Sidorov *et al.* (2002, 2020c) and Sidorov (2014b) outline a number of sections and applications of the modern theory of branching solutions of nonlinear equations with parameters. In this section, based on the methods outlined by Sidorov *et al.* (2020c), the existence theorems of bifurcation points for the solutions of Cauchy problems for differential equations with loadings given by the Stieltjes functions are proven. A method for constructing the solutions of such a problem in the neighbourhood of bifurcation points is presented.

4.3 Linear Fredholm Integral Equations with Functionals and Parameters

The theory of linear Fredholm integral-functional equations of the second kind with linear functionals and parameters is considered. The necessary and sufficient conditions are obtained for the coefficients of the equation and those parameter values in the neighbourhood for which the equation has solutions. The leading terms of the asymptotics of the solutions are constructed. The constructive method is proposed for constructing a solution in both the regular and irregular cases. In the regular case, the solution is constructed as a Taylor series in powers of the parameter, whereas in the irregular case, it is constructed as a Laurent series in powers of the parameter. An example is used to illustrate the proposed constructive theory and method.

This section deals with the novel theory of linear integral equations with linear functionals. Modern views on the fundamental laws of nature are often stated in terms of integral equations. Many inverse problems in mathematical physics can be formulated or reduced to non-classical integral equations. Dreglea and Sidorov (2018) reduced the problem of identification of external force and heat source density dynamics into one of solving the Volterra integral equations of the first kind. The analysis of integral operators includes questions of finding eigenvalues and adjoint functions, studying the convergence of their asymptotics, and the existence and convergence theorems of approximate methods (Sidorov *et al.*, 2020c; Sidorov, 2014b). At the end of 20th century, Khromov found a new class of integral operators with discontinuous kernels and began a systematic study of them (Khromov, 2006). Under highly general assumptions, he derived the conditions under which the eigenfunction expansions of these operators behave like trigonometric Fourier series. However, these conditions, as well as the construction of the classical discontinuous Fredholm resolvent in the form of the ratio of two integer analytic expansions over a parameter, are difficult to verify. Here, readers may refer to the books by Sidorov (2013a, 2014b) for a class of equations with discontinuous kernels.

Sidorov and Sidorov (2022) studied the branching solutions of the Cauchy problem for nonlinear loaded differential equations with bifurcation parameters. The purpose of this study was to prove the properties of the resolvent integral operator as applied to the Fredholm integral equations of the second kind with local and integral loads and to formulate and prove constructive theorems of existence and convergence to arrive at the desired solution of successive approximations.

Let us consider the equation

$$x - \mathcal{L}x - \lambda \mathcal{K}x = f, \tag{4.29}$$

where the linear operators $\mathcal{L}$ and $\mathcal{K}$ are given as follows:

$$\mathcal{L}x := \sum_{k=1}^{n} a_k(t) \langle \gamma_k, x \rangle,$$

$$\mathcal{K}x := \int_a^b K(t,s) x(s) \, ds.$$

λ is the parameter. All the functions in equation (4.29) are assumed to be continuous. The kernel $K(t,s)$ can be symmetric, and it is also continuous both in t and s. The desired solution $x(t)$ is constructed in $\mathcal{C}_{[a,b]}$.

The linear functionals $\langle \gamma_k, x \rangle$ in applications correspond to the *loads* imposed on the desired solution. The loads can be local, such as $\langle \gamma_k, x \rangle = x(t_k)$, $t_k \in [a,b]$, or integral, such as $\langle \gamma_k, x \rangle = \int_a^b \gamma_k(t) x(t) \, dt$, where $\gamma_k(t)$ are piecewise continuous functions for $t \in [a,b]$ or $\langle \gamma_k, x \rangle = \int_a^b x(t) \, d\gamma_k(t)$, where $\gamma_k(t)$ is a given function of limited variation.

The objective is to construct the solution $x(t, \lambda)$ for $\lambda \in \mathbb{R}^1$ for equation (4.29). For the operator $\mathcal{L}x$, the following is used as a brief notation:

$$\mathcal{L}x := \sum_{k=1}^{n} a_k(t) \langle \gamma_k, x \rangle \equiv (\boldsymbol{a}(t), \langle \boldsymbol{\gamma}, x \rangle),$$

where the conventional notation $(\cdot, \cdot)$ is used for the scalar product. Here, $\boldsymbol{a}(t) = (a_1(t), \ldots, a_n(t))^T$, $a_i(y) \in \mathcal{C}_{[a,b]}$, and $\langle \boldsymbol{\gamma}, x \rangle = (\langle \gamma_1, x \rangle, \ldots, \langle \gamma_n, x \rangle)^T$.

Loaded differential equations have been intensively studied during the past decades. The term *loaded equation* was first used in the works of Nakhushev. For more details, readers may refer to his monograph (2012). Loaded equations appear in many applications; see, for example, Chadam *et al.* (1992) and Baltaeva *et al.* (2022). However, theory and numerical methods for the loaded integral equations remained less developed. In a paper by Sidorov and Sidorov (2021), the problem statement for an integral equation with a single load is given. Then, Sidorov and Dreglea Sidorov (2022, 2023) proposed a theory of the Hammerstein integral equations with loads and bifurcation parameters. Lample and Rosenwasser (2010) employed the Fredholm resolvent for computing the H_2-norm of linear periodic systems.

A similar statement is addressed in the current section, followed by the description of an analytical method which makes it possible to consider integral equations with an arbitrary finite number of local and integral loads. An example of functionals that generate local and integral loads in the space $\mathcal{C}_{[a,b]}$ is the functional

$$\langle \gamma, x \rangle := \sum_{i=1}^{m} \alpha_i x(t_i) + \sum_{i=1}^{n} \int_{a_i}^{b_i} m_i(s) x(s) \, ds,$$

where $\alpha_i \in \mathbb{R}^1$, $[a_i, b_i] \subset [a, b]$, $m_i(s) \in \mathcal{C}_{[a_i, b_i]}$, $t_i \in [a, b]$.

4.3.1 *System of equations to determine the load*

Let us introduce the following condition:

I. $\langle \gamma_k, K(t,s) \rangle = 0$, $k = 1, \ldots, n$, $s \in [a, b]$ and the vectors $\boldsymbol{x}_\gamma = (\langle \gamma_1, x \rangle, \ldots, \langle \gamma_n, x \rangle)^T$, $\boldsymbol{f}_\gamma = (\langle \gamma_1, f \rangle, \ldots, \langle \gamma_n, f \rangle)^T$.

Lemma 4.3.1. *Let condition* **I** *be fulfilled. Then, the load vector* $\boldsymbol{x}_j$ *necessarily satisfies the system*

$$(E - A_0)\boldsymbol{c} = \boldsymbol{f}_\gamma, \tag{4.30}$$

where $A_0 = [\langle \gamma_i, a_k \rangle]_{i,k=1}^n$ *and* E *is an* $(n \times n)$ *identity matrix.*

Proof. Let us apply the functionals $\langle \gamma_i, \cdot \rangle$, $i = 1, \ldots, n$, to both parts of equation (4.29). Using **I**, the following system can be derived:

$$\langle \gamma_i, x \rangle - \sum_{k=1}^{n} \langle \gamma_i, a_k \rangle \langle \gamma_k, x \rangle = \langle \gamma_i, f \rangle, \ i = 1, \ldots, n. \tag{4.31}$$

The system of linear algebraic equations (4.31) is, in fact, system (4.30) presented in the coordinate system. The lemma is proved. $\qquad\square$

From this Lemma, we have the following.

Corollary 4.3.2. *Let condition* **I** *be fulfilled and system* (4.30) *have no solution. Then, equation* (4.29) *has no solution in the class of continuous functions.*

Let condition **I** be fulfilled and vector $c^* \in \mathbb{R}^n$ satisfy system (4.30). Then, the solution $x(t, \lambda)$ of equation (4.29) depends on the vector c^* and satisfies the following Fredholm integral equation of the second kind:

$$x(t, \lambda) - \lambda \int_a^b K(t, s) x(s, \lambda)\, ds = f(t) + (a(t), c^*).$$

Lemma 4.3.3. *The solution of equation* (4.29) *for arbitrary* λ, *except the characteristic numbers* λ_i *of the kernel* $K(t, s)$, *is defined by the following formula:*

$$x(t, \lambda) = (a(t), x_\gamma(\lambda)) + \int_a^b \Gamma(t, s, \lambda)(a(s), x_\gamma(\lambda))\, ds$$

$$+ \int_a^b \Gamma(t, s, \lambda) f(s)\, ds + f(t). \tag{4.32}$$

Here, $\Gamma(t, s, \lambda) = \frac{D(t, s, \lambda)}{D(\lambda)}$, *where* $D(t, s, \lambda)$ *and* $D(\lambda)$ *are entire analytic functions of parameter* λ, *with* $D(\lambda_i) = 0$. *The load vector* $x_\gamma(\lambda)$ *must necessarily satisfy the following system of* n *linear algebraic equations:*

$$(E - A_0 - A(\lambda)) x_\gamma(\lambda) = b(\lambda) \tag{4.33}$$

with matrix

$$A(\lambda) = \left\langle \gamma_i, \int_a^b \Gamma(t, s, \lambda) a_k(s)\, ds \right\rangle_{i,k=1}^n \tag{4.34}$$

and vector

$$b(\lambda) = \left\langle \gamma_i, f(t) + \int_a^b \Gamma(t, s, \lambda) f(s)\, ds \right\rangle_{i=1}^n.$$

The set of characteristic numbers $\{\lambda_i\}$ *is a finite and countable set.*

Proof. It is known (see Section 9(3) in Kolmogorov and Fomin, 1999) that an inverse operator $(I - \lambda K)^{-1}$ is defined by the Fredholm (1900) formula:

$$(I - \lambda K)^{-1} = I + \lambda \int_a^b \frac{D(t, s, \lambda)}{D(\lambda)} [\cdot] \, ds.$$

The functions $D(t, s, \lambda)$ and $D(\lambda)$ are entire analytical functions with respect to λ, defined for $\lambda \in \mathbb{R}^1$. Moreover, the characteristic numbers of the kernel $K(t, s)$ of the operator $\mathcal{K}$ are zeros of the denominator $D(\lambda)$. Thus, the inverse operator $(I - \lambda K)^{-1}$ can be called a *discontinuous operator*. Indeed, the function $\Gamma(t, s, \lambda)$ in solution (4.32) has the second-kind discontinuities at the points $\{\lambda_i\}$. By solving system (4.33) and substituting its solution into (4.32), we find the solution of the original problem (4.29). The lemma is proved. □

Remark 4.3.4. In system (4.33), in the general case, the matrix $A(\lambda)$ and the vector $\boldsymbol{b}(\lambda)$ will have second-kind discontinuities at points λ.

Let us distinguish the class of kernels $K(t, s)$ when the matrix A_0 and the vector $\boldsymbol{b}(\lambda)$ are specified. Let the kernel $K(t, s)$ generate the nilpotency of the operator $\mathcal{K}$.

Let $|\lambda| < \frac{1}{\|\mathcal{K}\|}$. In that case, the solution of the equation $x - \lambda \mathcal{K} x = f$ for an arbitrary source function f is defined uniquely as follows:

$$x = f + \lambda \mathcal{K} f + \lambda^2 \mathcal{K}^2 f + \cdots + \lambda^p \mathcal{K}^p f.$$

Here,

$$\mathcal{K}^n f = \int_a^b K_n(t, s) f(s) \, ds,$$

where

$$K_n(t, s) = \int_a^b K(t, z) K_{n-1}(z, s) \, dz.$$

Here, $K_1(t, s) := K(t, s)$, $K_{p+1}(t, s) = 0$ due to the nilpotency of the operator $\mathcal{K}$ for some $p \geq 1$. Therefore, formula (4.32) can be presented in the following constructive form:

$$x(t, \lambda, \boldsymbol{x}_\gamma) = f(t) + (\boldsymbol{a}(t), \boldsymbol{x}_\gamma) + \int_a^b \left(\lambda K(t, s) + \lambda^2 K_2(t, s) + \cdots \right.$$

$$\cdots + \lambda^p K_p(t, s) \Big) (f(s) + (\boldsymbol{a}(s), \boldsymbol{x}_\gamma)) \, ds. \qquad (4.35)$$

Correspondingly, we derive the refined system of linear algebraic equations (4.33) with respect to the load vector because

$$A(\lambda) = \left\langle \gamma_i, \int_a^b (\lambda K(t, s) + \lambda^2 K_2(t, s) + \cdots + \lambda^p K_p(t, s)) a_k(s) \, ds \right\rangle_{i,k=1}^n , \qquad (4.36)$$

$$\boldsymbol{b}(\lambda) = \left\langle \gamma_i, \ f(t) + \int_a^b (\lambda K(t, s) + \lambda^2 K_2(t, s) + \cdots + \lambda^p K_p(t, s)) f(s) \, ds \right\rangle_{i=1}^n . \qquad (4.37)$$

Thus, $A(\lambda)$ and $\boldsymbol{b}(\lambda)$ are continuous in λ. It should be noted that if $\langle \gamma_i, K(t, s) \rangle = 0$, $i = 1, \ldots, n$, then $A(\lambda) = 0$, and system (4.33) degenerates into system (4.30). Therefore, in this case, the vector $\boldsymbol{x}_\gamma$ from solution (4.35) to the given problem (4.29) can be determined. Then, the following theorem can be formulated.

Theorem 4.3.5. *Let the operator $\mathcal{K}$ be nilpotent and $\langle \gamma_i, K(t, s) \rangle = 0$, $i = 1, \ldots, n$, $\forall s \in [a, b]$. Then, a solution to equation (4.29) exists as the functional polynomial (4.35) of pth order in the parameter λ. The coefficients of polynomial (4.35) depend on the selection of the load vector $\boldsymbol{x}_\gamma$ in $\mathbb{R}^n$.*

If the operator $\mathcal{K}$ is not nilpotent and the identity $\langle \gamma_i, K(t, s) \rangle = 0$ is not satisfied, then the solution $x(t, \lambda)$ of equation (4.29) can be found in the class of functions continuous in t. This solution can be represented in the punctured neighborhood $0 < |\lambda| < \rho$ in the form of a Laurent series with pole at the point $\lambda = 0$.

4.3.2   *Successive approximations*

Let $\det(E - A_0) \neq 0$. Then, there exists a neighborhood of λ $|\lambda| < \rho$ such that system (4.33) has a solution, $\boldsymbol{x}_\rho(\lambda) \to (E - A_0)^{-1} \boldsymbol{f}_\gamma$, as $\lambda \to 0$. A positive ρ exists since $\|(E - A_0)^{-1} A(\lambda)\| \to 0$ as $\lambda \to 0$.

Let us call the case of $\det(E - A_0) \neq 0$ *regular*.

Theorem 4.3.6. *In the regular case of* $\det(E - A_0) \neq 0$, *there exists a neighborhood* $|\lambda| < \rho$ *in which equation* (4.29) *has a unique solution continuous in t and holomorphic in* λ.

Corollary 4.3.7. *Let* $\det(E - A_0) \neq 0$, $\|(I - L)^{-1}\mathcal{K}\| \leq l$. *Fix the scalar as* $q < 1$. *Then, for* $|\lambda| \leq \frac{q}{l}$, *equation* (4.29) *has a unique solution. Moreover, the solution is holomorphic in* λ. *The sequence* $\{x_n\}$, *where* $x_n = \lambda(I - L)^{-1}\mathcal{K}x_{n-1}+(I-L)^{-1}f$, $x_0 = 0$, *uniformly converges to the desired solution* $x(t, \lambda)$ *of equation* (4.29) *at the rate of a geometric progression with the denominator* $q < 1$.

Let us focus now on the irregular case of $\det(E - A_0) = 0$. Let $A_0 = E$. Then, $\det(E - A_0) = 0$, and we have an irregular case. Let $\frac{d^i}{d\lambda^i}A(\lambda)\big|_{\lambda=0}$, for $i = 0, 1, \ldots, p - 1$, be zero matrices and $\frac{d^p}{d\lambda^p}A(\lambda)\big|_{\lambda=0} \neq 0$. Then, the load vector x_γ satisfies the following system:

$$\left(-E - A_p^{-1} \sum_{m=p+1}^{\infty} \lambda^{m-p}A_m\right)x_\gamma = \lambda^{-p}A^{-1}b(\lambda),$$

where

$$A_p = \frac{1}{p!}\left(\frac{d^p}{d\lambda^p}A(\lambda)\right)\bigg|_{\lambda=0}.$$

Let's select the neighborhood $|\lambda| < \rho$ such that

$$\left\|A_p^{-1} \sum_{m=p+1}^{\infty} \lambda^{m-p}A_m\right\| \leq q < 1.$$

Then,

$$\lambda^p x_\gamma = -\sum_{n=0}^{\infty}\left(-A_p^{-1}\sum_{m=p}^{\infty}\lambda^{m-p}A_m\right)^n A_p^{-1}b(\lambda),$$

which is a series that converges to the holomorphic function

$$\nu(\lambda) = -\sum_{n=0}^{\infty}\left(-A_p^{-1}\sum_{m=p}^{\infty}\lambda^{m-p}A_m\right)^n A_p^{-1}b(\lambda)$$

at the rate of a geometric sequence with the denominator $q < 1$ for $|\lambda| \leq \rho$. Therefore, the load $x_\gamma(\lambda) = \lambda^{-p}\nu(\lambda)$ is a Laurent series with a pth-order pole. Then, the following theorem is true.

Theorem 4.3.8. *Let* $A_0 = E$, $A(\lambda) = \sum_{m=p}^{\infty}A_m\lambda^m$, $p \geq 1$. *Let the matrix* A_p *be non-singular. Then, there exists a punctured neighborhood,*

$0 < |\lambda| \leq p$, *such that equation* (4.29) *has a solution*, $x(t, \lambda)$, *with a pole at point* $\lambda = 0$ *of order less than or equal to* p.

Let us consider the equation

$$x(t, \lambda) - a(t)x(0, \lambda) = \lambda \int_0^1 b(t)m(s)x(s, \lambda)\, ds + f(t), \quad t \in [0, 1].$$

Let us have the irregular case of $a(0) = 1$. Let $b(0) \neq 0$, i.e. condition **I** is not fulfilled:

$$\frac{d}{d\lambda} A(\lambda)\big|_{\lambda=0} = b(0) \int_0^1 m(s)a(s)\, ds.$$

Let

$$\int_0^1 m(s)a(s)\, ds \neq 0.$$

Then, all the conditions in Theorem 4.3.8 are fulfilled for $p = 1$. Then, the equation has the solution $x(\lambda)$ for $|\lambda| > 0$ with a first-order pole at the point $\lambda = 0$. The desired solution is the following:

$$x(t, \lambda) = f(t, \lambda) - \frac{b(t)}{b(0)} f(0) + a(t)x(0, \lambda),$$

where the load $x(0, \lambda)$ is constructed as follows:

$$x(0, \lambda) \equiv \frac{1}{(a, m)} \left[-\frac{f(0)}{\lambda b(0)} - (f, m) + \frac{f(0)}{b(0)} (b, m) \right],$$

where $(a, m) = \int_0^1 a(t)m(t)dt$, $(f, m) = \int_0^1 f(t)m(t)dt$, and $(b, m) = \int_0^1 b(t)m(t)dt$. In this example, the solution is constructed in an explicit form.

To summarise, the linear Fredholm integral functional equations of the second kind with linear functionals are studied in this section. Necessary and sufficient conditions are formulated. Constructive methods are proposed for both regular and irregular cases. The solution in the form of a Taylor series is constructed in terms of powers of the parameters. In the irregular case, the solution is constructed as a Laurent series of powers of the parameters. The constructive theory and methods are demonstrated using a model example. However, the case of $A_0 \neq E$ remains unaddressed in this monograph. The most complete results can be derived for the case of the symmetric matrix A_0. In that case, the solution of equation (4.29) can also be presented as a Laurent series with a pole at the point $\lambda = 0$. The corresponding sufficient condition can be derived based on generalised Jordan chains of the theory of perturbed nonlinear operators (Vainberg and Trenogin, 1964). The bifurcation theory of nonlinear loaded integral equations, using this approach in combination with representation theory and group symmetry (Loginov and Sidorov, 1992), will also be addressed in future works. Some results in this direction have been reported by Sidorov and Dreglea Sidorov (2023), Sidorov *et al.* (2020c), and Sidorov (2014b).

Chapter 5

Applications in Electrical Engineering and Automation

5.1 Boundary Value Problems in Power Electronics: Non-insulated Magnetic Regime in a Vacuum Diode

In this section, we study a stationary boundary value problem (BVP) derived from the magnetic insulating (or non-insulating) regime in a planar diode. The main goal is to prove the existence of non-negative solutions for this nonlinear singular system of second-order ordinary differential equations. To achieve this goal, we reduce the BVP to a singular system of coupled nonlinear Fredholm integral equations and then analyse its solvability by investigating the existence of fixed points in the related operators. This system of integral equations is studied by means of the Leray–Schauder topological degree theory.

High-energy devices, such as vacuum diodes, are designed to operate at extremely high applied voltages. Vacuum diodes play a crucial role in high-voltage direct current (HVDC) transmission systems. In HVDC systems, vacuum diodes are used as rectifiers to convert alternating current (AC) into direct current (DC). They allow current to flow in only one direction, making them essential for converting the power generated in AC form to DC for efficient transmission over long distances. In addition to their role in rectification, vacuum diodes also provide high-voltage capabilities and low power losses. Overall, vacuum diodes are important components of HVDC systems and are essential for their efficient and reliable operation.

The saturation of current due to self-consistent electric and magnetic fields is a nonlinear phenomenon under electron transport. Langmuir and Compton (1931) initiated investigations into this phenomenon, establishing explicit formulas for the saturation current in both planar and symmetric diode cases. They assumed that the saturation current reaches its maximum value when the electric field vanishes at the emission cathode. This condition is referred to as the Child–Langmuir condition, and the diode is said to operate under a space-charge-limited, or Child–Langmuir, regime.

Here, two basic regimes are possible: the first is when electrons reach the anode, i.e. the 'non-insulated diode', and the second is when, due to the extremely high electric and magnetic fields applied, electrons rotate back to the cathode—the so-called 'insulated diode'. In the latter case, there is an electronic layer beyond which the electromagnetic field equals zero.

The non-insulated diode regime is described by the following nonlinear two-point coupled second-order BVP:

$$\frac{d^2\varphi(x)}{dx^2} = j_x \frac{1+\varphi(x)}{\sqrt{(1+\varphi(x))^2 - 1 - a(x)^2}}, \quad \varphi(0) = 0, \quad \varphi(1) = \varphi_L,$$

$$\frac{d^2 a(x)}{dx^2} = j_x \frac{a(x)}{\sqrt{(1+\varphi(x))^2 - 1 - a(x)^2}}, \quad a(0) = 0, \quad a(1) = a_L, \tag{I}$$

where $j_x > 0$ is a constant (independent of x), φ is the potential of electric field, a is the potential of magnetic field, and $x \in [0, 1]$. For details about the problem setting and the complete derivation of system (I), see, for example, Ben Abdallah *et al.* (1998).

The existence of solutions for system (I) was studied by Ben Abdallah *et al.* (1987) using a shooting method, with $\beta = a'(0)$ and j_x as shooting parameters. The strategy involved the following steps: given the values of β and j_x, solve (I) with the Cauchy conditions $\varphi(0) = 0$, $a(0) = 0$, $\varphi'(0) = 0$, and $a'(0) = \beta$, and then adjust the values in order to fulfill the conditions $\varphi(1) = \varphi_L$ and $a(1) = a_L$.

In this section, we analyse the existence of (non-negative) solutions for the coupled second-order BVP (I) by first transforming it into a coupled system of singular nonlinear Fredholm integral equations. Then, we investigate the conditions to ensure the non-negativity of the image functions resulting from the evaluation of these integral equations. Finally, the existence of such solutions is guaranteed by using the classical Leray–Schauder topological degree theory.

We recall that some analytical and numerical aspects of the Child–Langmuir regime were recently reported by Degond *et al.* (1987) and by Dulov and Sinitsyn (2005). Also, we would like to point out that a number of interesting results concerning the theory and application of functional group methods under model symmetry conditions and the bifurcation of desired solutions are presented in the monographs by Sidorov *et al.* (2002, 2020c).

5.1.1   *Non-negative solutions for BVPs*

Now, we are interested in investigating the existence of non-negative solutions for a BVP derived from the non-insulated regime for a vacuum planar diode. After the substitutions $\varphi + 1 =: u$ and $a =: v$, BVP (I) reads

$$u''(x) = j_x \frac{u(x)}{\sqrt{u^2(x) - 1 - v^2(x)}}, \quad u(0) = 1, \ u(1) = \varphi_L + 1 = p,$$

$$v''(x) = j_x \frac{v(x)}{\sqrt{u^2(x) - 1 - v^2(x)}}, \quad v(0) = 0, \ v(1) = a_L = q. \tag{5.1}$$

By *singular*, we mean that $u^2(x) = 1 + v^2(x)$ is allowed. First, we reduce this system of BVPs to a coupled system of integral equations. Through integration by parts and by using the boundary conditions for the second-order BVP,

$$u''(x) = j_x \frac{u(x)}{\sqrt{u^2(x) - 1 - v^2(x)}}, \quad u(0) = 1, \ u(1) = p,$$

we have

$$u'(x) = u'(0) + j_x \int_0^x \frac{u(s)}{\sqrt{u^2(s) - 1 - v^2(s)}} ds.$$

Hence, integrating one more time, we obtain

$$u(x) = 1 + u'(0)x + j_x \int_0^x \frac{(x - s)u(s)}{\sqrt{u^2(s) - 1 - v^2(s)}} ds.$$

Using the boundary condition $u(1) = p$, we get

$$u'(0) = p - 1 - j_x \int_0^1 \frac{(1 - s)u(s)}{\sqrt{u^2(s) - 1 - v^2(s)}} ds.$$

Then, we reduce the second-order BVP to the following integral equation:

$$u(x) = 1 + (p-1)x - j_x \cdot x \int_0^1 \frac{(1-s)u(s)}{\sqrt{u^2(s) - 1 - v^2(s)}} ds$$

$$+ j_x \int_0^x \frac{(x-s)u(s)}{\sqrt{u^2(s) - 1 - v^2(s)}} ds$$

$$= 1 + (p-1)x + j_x \int_0^1 \frac{x(s-1)u(s) + \chi_{(0,x)}(s)(x-s)u(s)}{\sqrt{u^2(s) - 1 - v^2(s)}} ds$$

$$= 1 + (p-1)x + j_x \int_0^1 \frac{G(s,x)u(s)}{\sqrt{u^2(s) - 1 - v^2(s)}} ds,$$

where $\chi_{(0,x)}(s)$ is the characteristic function of the interval $(0, x)$ and

$$G(s,x) = \begin{cases} s(x-1), & 0 \le s \le x \le 1, \\ x(s-1), & 0 \le x \le s \le 1. \end{cases}$$

In the same fashion, we can reduce the second-order BVP

$$v''(x) = j_x \frac{v(x)}{\sqrt{u^2(x) - 1 - v^2(x)}}, \quad v(0) = 0, \ v(1) = q,$$

to the integral equation

$$v(x) = qx - j_x \cdot x \int_0^1 \frac{(1-s)v(s)}{\sqrt{u^2(s) - 1 - v^2(s)}} ds + j_x \int_0^x \frac{(x-s)v(s)}{\sqrt{u^2(s) - 1 - v^2(s)}} ds$$

$$= qx + j_x \int_0^1 \frac{G(s,x)v(s)}{\sqrt{u^2(s) - 1 - v^2(s)}} ds.$$

In this way, analysing the existence of solutions for system (5.1) is equivalent to investigating the existence of solutions for the following coupled system of nonlinear singular Fredholm integral equations:

$$u(x) = 1 + (p-1)x + j_x \int_0^1 \frac{G(s,x)u(s)}{\sqrt{u^2(s) - 1 - v^2(s)}} ds,$$

$$v(x) = qx + j_x \int_0^1 \frac{G(s,x)v(s)}{\sqrt{u^2(s) - 1 - v^2(s)}} ds.$$

$$(5.2)$$

The equations are posed in the Banach space $X = (C^1[0,1], \mathbb{R})$ endowed with the norm $\|f\|_\infty = \max\{|f(x)| : x \in [0,1]\}$. By $X \times X$, we denote the product space, which is a Banach space under the norm $\|(u,v)\| = \max\{\|u\|_\infty, \|v\|_\infty\}$, and we define the operator $F(u,v) : X \times X \longrightarrow X \times X$ by the formula

$$F(u,v) = (F_1(u,v), F_2(u,v)),$$

where, for each $x \in [0,1]$,

$$F_1(u,v)(x) = 1 + (p-1)x + j_x \int_0^1 \frac{G(s,x)u(s)}{\sqrt{u^2(s) - 1 - v^2(s)}} ds,$$

$$F_2(u,v)(x) = qx + j_x \int_0^1 \frac{G(s,x)v(s)}{\sqrt{u^2(s) - 1 - v^2(s)}} ds.$$

Note that for $u, v \in X$, the mappings $F_1(u,v)$ and $F_2(u,v)$ are well defined if $\sqrt{u^2(s) - 1 - v^2(s)} \in \mathbb{R}$, so we require that

$$u^2(s) \geq 1 + v^2(s), \quad \text{for all } s \in [0,1]. \tag{5.3}$$

In particular, $|u(s)| \geq 1$ for all $s \in [0,1]$, but $|u| \not\equiv 1$. Moreover, we have the following inequality.

Lemma 5.1.1. *Under condition* (5.3)*, the inequality*

$$|uu' - vv'| \geq 1 \tag{5.4}$$

holds for all $s \in [0,1]$.

Proof. We should prove that either $uu' - vv' \geq 1$ or $uu' - vv' \leq -1$. If we assume that $uu' - vv' < 1$, then by integrating the inequality above, we have $u^2(s)/2 < s + v^2(s)/2 + K$ for any constant K; however, from (5.3), we have

$$\frac{1}{2} + \frac{v^2(s)}{2} < s + \frac{v^2(s)}{2} + K,$$

i.e. $1/2 < s + K$ for any K, which is false. Take, for instance, $K = -1/2$. Similarly, if $uu' - vv' > -1$, by integrating, we get $u^2(s)/2 > -s + v^2(s)/2 + C$ for any $C \in \mathbb{R}$ and all $s \in [0,1]$. But this is not necessarily true for constants $C < 1/2$. $\qquad \square$

In the operator theory scheme, the existence of a solution for system (5.2) is equivalent to the existence of a fixed point for the operator $F(u, v)$ on $X \times X$. Since we want to find solutions for system (5.1) that is physically meaningful, we are interested in the existence of non-negative solutions for this system. Thus, we assume that the boundary conditions satisfy $p \geq 1$ and $q \geq 0$.

Let P be the cone of all non-negative functions on X, i.e.

$$P = \{f \in C^1([0, 1], \mathbb{R}) \ : \ f(x) \geq 0, \text{ for all } x \in [0, 1]\}.$$

Our first step is to find the conditions under which $F(u, v) : X \times X \longrightarrow P \times P$, that is, for each $x \in [0, 1]$, the following inequalities are satisfied:

$$F_1(u, v)(x) = 1 + (p - 1)x + j_x \int_0^1 \frac{G(s, x)u(s)}{\sqrt{u^2(s) - 1 - v^2(s)}} ds \geq 0, \qquad (5.5)$$

$$F_2(u, v)(x) = qx + j_x \int_0^1 \frac{G(s, x)v(s)}{\sqrt{u^2(s) - 1 - v^2(s)}} ds \geq 0. \qquad (5.6)$$

Equivalently, inequalities (5.5) and (5.6) are satisfied, for all $x \in [0, 1]$, if the following inequalities hold, respectively:

$$-j_x \int_0^1 \frac{G(s, x)u(s)}{\sqrt{u^2(s) - 1 - v^2(s)}} ds \leq 1 + (p - 1)x, \qquad (5.7)$$

$$-j_x \int_0^1 \frac{G(s, x)v(s)}{\sqrt{u^2(s) - 1 - v^2(s)}} ds \leq qx. \qquad (5.8)$$

Note that $0 \leq -G(s, x) \leq 1$, for all $s, x \in [0, 1]$. Thus, for any $f \in X$,

$$-j_x \int_0^1 \frac{G(s, x)f(s)}{\sqrt{u^2(s) - 1 - v^2(s)}} ds \leq j_x \int_0^1 \frac{f(s)}{\sqrt{u^2(s) - 1 - v^2(s)}} ds$$

$$\leq \left| j_x \int_0^1 \frac{u(s)u'(s) - v(s)v'(s)}{\sqrt{u^2(s) - 1 - v^2(s)}} \frac{f(s)}{u(s)u'(s) - v(s)v'(s)} ds \right|$$

$$\leq j_x \int_0^1 \frac{\left| \left(\sqrt{u^2(s) - 1 - v^2(s)} \right)' \right| |f(s)|}{|u(s)u'(s) - v(s)v'(s)|} ds.$$

From Hölder's inequality, we have

$$\int_0^1 \frac{\left|\left(\sqrt{u^2(s)-1-v^2(s)}\right)'\right| |f(s)|}{|u(s)u'(s)-v(s)v'(s)|}ds \leq \int_0^1 \left|\left(\sqrt{u^2(s)-1-v^2(s)}\right)'\right| ds$$

$$\times \left\|\frac{|f|}{|uu'-vv'|}\right\|_\infty$$

$$\leq \sqrt{p^2-1-q^2}\left\|\frac{|f|}{|uu'-vv'|}\right\|_\infty .$$

On the other hand, inequality (5.4) implies

$$\left\|\frac{|f|}{|uu'-vv'|}\right\|_\infty \leq \|f\|_\infty.$$

Hence, we obtain the following estimate:

$$-j_x \int_0^1 \frac{G(s,x)f(s)}{\sqrt{u^2(s)-1-v^2(s)}}ds \leq j_x\sqrt{p^2-1-q^2}\|f\|_\infty, \quad \text{for all } x \in [0,1].$$

$$(5.9)$$

Therefore, inequality (5.7), and so inequality (5.5), holds if

$$j_x\sqrt{p^2-1-q^2}\|u\|_\infty \leq 1.$$

That is, inequality (5.5) holds for any $u \in X$ satisfying

$$\|u\|_\infty \leq \frac{1}{j_x\sqrt{p^2-1-q^2}}.$$

Now, we find the conditions under which inequality (5.8) (and consequently, inequality (5.6)) is satisfied:

$$-j_x \int_0^1 \frac{G(s,x)v(s)}{\sqrt{u^2(s)-1-v^2(s)}}ds = j_x \int_0^1 \frac{x(1-s)v(s)}{\sqrt{u^2(s)-1-v^2(s)}}ds$$

$$+ j_x \int_0^x \frac{(s-x)v(s)}{\sqrt{u^2(s)-1-v^2(s)}}ds$$

$$\leq j_x \left|\int_0^1 \frac{x(1-s)v(s)}{\sqrt{u^2(s)-1-v^2(s)}}ds\right|$$

$$+ j_x \left|\int_0^x \frac{(s-x)v(s)}{\sqrt{u^2(s)-1-v^2(s)}}ds\right|.$$

Note that

$$\max_{s\in[0,1]}\{x(1-s)\} = x \quad \text{and} \quad s - x \le 0.$$

Then, $s - x \le x$ for all $x \in [0,1]$. These facts and estimate (5.9) give

$$-j_x \int_0^1 \frac{G(s,x)v(s)}{\sqrt{u^2(s)-1-v^2(s)}} ds \le j_x \int_0^1 \frac{x|v(s)|}{\sqrt{u^2(s)-1-v^2(s)}} ds$$

$$+ j_x \int_0^x \frac{x|v(s)|}{\sqrt{u^2(s)-1-v^2(s)}} ds$$

$$\le j_x \int_0^1 \frac{x|v(s)|}{\sqrt{u^2(s)-1-v^2(s)}} ds$$

$$+ j_x \int_0^1 \frac{x|v(s)|}{\sqrt{u^2(s)-1-v^2(s)}} ds$$

$$\le 2xj_x\sqrt{p^2-1-q^2}\|v\|_\infty.$$

In this way, inequality (5.8) holds, for all $x \in [0,1]$, if

$$2xj_x\sqrt{p^2-1-q^2}\|v\|_\infty \le qx,$$

i.e. $F_2(u,v)(x) \ge 0$, for all $v \in X$, satisfying

$$\|v\|_\infty \le \frac{q}{2j_x\sqrt{p^2-1-q^2}}.$$

Let us denote by $\overline{\Omega}_1 := \overline{B}\left(\frac{1}{j_x\sqrt{p^2-1-q^2}}\right)$ the closed ball in X centered at 0 with radii $1/j_x\sqrt{p^2-1-q^2}$, and let $\overline{\Omega}_2 := \overline{B}\left(\frac{q}{2j_x\sqrt{p^2-1-q^2}}\right)$ denote a closed ball in X centered at 0 with radii $q/2j_x\sqrt{p^2-1-q^2}$.

We just proved the following statement.

Proposition 5.1.2. *The operator $F(u,v)$ applies $\overline{\Omega}_1 \times \overline{\Omega}_2$ into $P \times P$.*

Remark 5.1.3. Note that from the triangle inequality and estimate (5.9), in $\overline{\Omega}_1 \times \overline{\Omega}_2 \subset X \times X$, we get

$$\|F_1(u,v)\|_\infty \le p+1,$$

$$\|F_2(u,v)\|_\infty \le 2q.$$

Hence, the operator $F(u, v)$ satisfies

$$\|F(u, v)\| \leq \max\{p + 1, 2q\}.$$

To show the existence of at least one fixed point for the operator $F(u, v)$ on $\Omega_1 \times \Omega_2$ (which implies the existence of non-negative solutions for system (5.1)), we use the well-known Leray–Schauder topological degree theory (see, for example, Deimling, 2010). More precisely, we compute $d(\varphi, \Omega_1 \times \Omega_2, (0, 0))$, where φ is the compact perturbation of the identity given by $\varphi(u, v) = I_{X \times X}(u, v) - F(u, v)$. Here, $I_{X \times X}$ denotes the identity map of $X \times X$.

First, we should prove that $F(u, v)$ is, in fact, completely continuous.

Proposition 5.1.4. *The operator* $F(u, v) : \overline{\Omega}_1 \times \overline{\Omega}_2 \longrightarrow P \times P$ *is continuous and compact.*

Proof. Let $(u, v) \in \overline{\Omega}_1 \times \overline{\Omega}_2$. First, we prove the continuity of the operator $F(u, v)$, for which we prove that the mappings

$$F_1(u, v) : \overline{\Omega}_1 \times \overline{\Omega}_2 \longrightarrow P,$$

$$F_2(u, v) : \overline{\Omega}_1 \times \overline{\Omega}_2 \longrightarrow P$$

are continuous. Let $(u, v) \in \overline{\Omega}_1 \times \overline{\Omega}_2$ such that $(u_n, v_n) \rightrightarrows (u, v)$. We prove the continuity of $F_1(u, v)$. Proving the continuity of $F_2(u, v)$ is similar. Since

$$\frac{G(s, x)u_n(s)}{\sqrt{u_n^2(s) - 1 - v_n^2(s)}} \rightrightarrows \frac{G(s, x)u(s)}{\sqrt{u^2(s) - 1 - v^2(s)}},$$

we conclude that

$$\lim_{n \to \infty} F_1(u_n, v_n)(x) = \lim_{n \to \infty} j_x \int_0^1 \frac{G(s, x)u_n(s)}{\sqrt{u_n^2(s) - 1 - v_n^2(s)}} ds$$

$$= j_x \int_0^1 \lim_{n \to \infty} \frac{G(s, x)u_n(s)}{\sqrt{u_n^2(s) - 1 - v_n^2(s)}} ds$$

$$= \int_0^1 \frac{G(s, x)u(s)}{\sqrt{u^2(s) - 1 - v^2(s)}} ds = F_1(u, v)(x)$$

is uniform on $\overline{\Omega}_1 \times \overline{\Omega}_2$. Hence, $F_1(u, v)$ is continuous.

To prove the equicontinuity of $F(u, v) \left(\overline{\Omega}_1 \times \overline{\Omega}_2\right)$, let $z_n = (u_n, v_n)$ be a sequence in $\overline{\Omega}_1 \times \overline{\Omega}_2$ and $x_1 \leq x_2$. Then,

$$|F_1(u_n, v_n)(x_1) - F_1(u_n, v_n)(x_2)|$$

$$\leq (p - 1)|x_1 - x_2| + j_x \int_0^1 \frac{|G(s, x_1) - G(s, x_2)|u_n(s)}{\sqrt{u_n^2(s) - 1 - v_n^2(s)}} ds,$$

where

$$|G(s, x_1) - G(s, x_2)| = \begin{cases} |s(x_1 - 1) - s(x_2 - 1)|, & 0 \leq s \leq x_1 \leq x_2 \leq 1, \\ |x_1(s - 1) - x_2(s - 1)|, & 0 \leq x_1 \leq x_2 \leq s \leq 1, \\ |s(x_1 - 1) - x_2(s - 1)|, & 0 \leq x_1 \leq s \leq x_2 \leq 1 \end{cases}$$

$$= \begin{cases} |s(x_1 - x_2)|, & 0 \leq s \leq x_1 \leq x_2 \leq 1, \\ |(s - 1)(x_1 - x_2)|, & 0 \leq x_1 \leq x_2 \leq s \leq 1, \\ |s(x_1 - x_2) - s + x_2|, & 0 \leq x_1 \leq s \leq x_2 \leq 1. \end{cases}$$

In all these cases, $|G(s, x_1) - G(s, x_2)| \to 0$, as $|x_1 - x_2| \to 0$. Therefore,

$$|F_1(u_n, v_n)(x_1) - F_1(u_n, v_n)(x_2)| \to 0, \quad \text{as } |x_1 - x_2| \to 0.$$

Similarly, we have

$$|F_2(u_n, v_n)(x_1) - F_2(u_n, v_n)(x_2)|$$

$$\leq q|x_1 - x_2| + j_x \int_0^1 \frac{|G(s, x_1) - G(s, x_2)|v_n(s)}{\sqrt{u_n^2(s) - 1 - v_n^2(s)}} ds$$

$$\to 0, \quad \text{as } |x_1 - x_2| \to 0.$$

Thus, from the Arzelá–Ascoli theorem, the operator $F(u, v)$ is compact. $\square$

The existence of at least one non-negative solution for the coupled singular system (5.1) is proved in the following theorem.

Theorem 5.1.5. *BVP* (5.1) *has at least one non-negative solution* (u, v), *provided that*

$$j_x < \min \left\{ \frac{1}{(p + 1)\sqrt{p^2 - 1 - q^2}}, \frac{1}{4\sqrt{p^2 - 1 - q^2}} \right\}. \tag{5.10}$$

Proof. Consider a compact perturbation of the identity $\varphi(u,v) = I_{X\times X}(u,v) - F(u,v)$. We demonstrate an admissible homotopy between φ and $I_{X\times X}$. Let the map $H((u,v),t) \in C((\Omega_1 \times \Omega_2) \times [0,1], X \times X)$ be given by

$$H((u,v),t) = t\varphi(u,v) + (1-t)I_{X\times X}(u,v).$$

Note that

$$H((u,v),0) = I_{X\times X}(u,v) = (u,v),$$
$$H((u,v),1) = \varphi(u,v).$$

We now show that H never assumes the value $(0,0)$ on the boundary $\partial\Omega_1 \times \partial\Omega_2$ of the set $\Omega_1 \times \Omega_2$. In fact, for $t = 0$,

$$H((u,v),0) = (u,v) \neq (0,0) \quad \text{over } \partial\Omega_1 \times \partial\Omega_2.$$

For $t = 1$,

$$H((u,v),1) = \varphi(u,v) = (u - F_1(u,v), v - F_2(u,v)).$$

If $H((u,v),1) = (0,0)$, we have

$$u - F_1(u,v) = 0,$$
$$v - F_2(u,v) = 0.$$

The inequalities above imply

$$\|F_1(u,v)\|_\infty = \|u\|_\infty,$$
$$\|F_2(u,v)\|_\infty = \|v\|_\infty.$$

From Remark 5.1.3, we conclude that

$$p + 1 \geq \|F_1(u,v)\|_\infty = \frac{1}{j_x\sqrt{p^2 - 1 - q^2}},$$
$$2q \geq \|F_2(u,v)\|_\infty = \frac{q}{2j_x\sqrt{p^2 - 1 - q^2}}.$$

However, this contradicts (5.10); therefore, $H((u,v),1) \neq (0,0)$. Finally, let $t_0 \in (0,1)$. If $H((u,v),t_0) = (0,0)$, then

$$t_0\varphi(u,v) + (1-t_0)I_{X\times X}(u,v) = (0,0),$$
$$t_0(u,v) - t_0F(u,v) + (u,v) - t_0(u,v) = (0,0),$$
$$(u,v) - t_0F(u,v) = (0,0).$$

We have the following scalar equalities:

$$t_0 F_1(u, v) = u,$$

$$t_0 F_2(u, v) = v$$

which are equivalent to

$$t_0 \|F_1(u, v)\|_\infty = \|u\|_\infty,$$

$$t_0 \|F_2(u, v)\|_\infty = \|v\|_\infty.$$

Again, Remark 5.1.3 gives us

$$t_0(p+1) \geq \frac{1}{j_x \sqrt{p^2 - 1 - q^2}},$$

implying that

$$j_x \geq \frac{1}{t_0(p+1)\sqrt{p^2 - 1 - q^2}} \geq \frac{1}{(p+1)\sqrt{p^2 - 1 - q^2}}$$

and

$$2t_0 q \geq \frac{q}{2j_x\sqrt{p^2 - 1 - q^2}},$$

which, in turn, implies

$$j_x \geq \frac{1}{4t_0\sqrt{p^2 - 1 - q^2}} \geq \frac{1}{4\sqrt{p^2 - 1 - q^2}}.$$

The inequalities above contradict condition (5.10). So, H never assumes the value $(0,0)$ on $\partial\Omega_1 \times \partial\Omega_2$. Hence, H is an admissible homotopy between φ and $I_{X \times X}$. Therefore,

$$d(\varphi, \Omega_1 \times \Omega_2, (0,0)) = d(I_{X \times X}, \Omega_1 \times \Omega_2, (0,0)) = 1.$$

This means that the equation $(u, v) - F(u, v) = (0,0)$ has at least one solution on $\Omega_1 \times \Omega_2$. $\square$

Remark 5.1.6. Note that condition (5.10) can be rewritten as

$$j_x < \begin{cases} \dfrac{1}{4\sqrt{p^2 - 1 - q^2}} & \text{if } 1 \leq p < 3, \\[3mm] \dfrac{1}{(p+1)\sqrt{p^2 - 1 - q^2}} & \text{if } p \geq 3. \end{cases}$$

5.2 Mathematical Model of Anaerobic Digestion for Biogas Production Using the Abel Equation

Biomass gasifier systems are important components for sustainable development of the national economy. They make it possible to implement the rational utilisation of renewable biomass. Biomass gasifier systems convert organic materials such as wood, agricultural residues, and industrial waste into a combustible gas through a process called gasification. By using renewable biomass as a fuel source, these systems contribute to sustainable energy production. Moreover, biomass gasification provides an alternative to fossil fuels, reducing the combustion of non-renewable resources and decreasing carbon dioxide emissions, a major contributor to climate change. Additionally, the carbon emitted during the combustion of biomass is part of a natural carbon cycle, making it a more sustainable energy option. Through the use of agricultural residues, forestry by-products, and organic waste, biomass gasification enables the beneficial reuse of organic materials that might otherwise be discarded or left to decompose, potentially contributing to environmental pollution. This supports sustainable waste management practices and reduces the environmental impact of organic waste. Biomass gasifier systems serve as a key component of providing energy access for rural and remote areas in large countries, such as Russia, China, and India.

In regions with limited access to traditional energy sources, biomass gasifier systems can be an important source of decentralised, distributed energy generation. It is important to combine different sources, including photovoltaic, biomass, and diesel, to build flexible hybrid systems (Kozlov *et al.*, 2020; Tomin *et al.*, 2022).

It should be outlined that decentralised distributed energy generation can improve living conditions, support economic activities, and enhance the quality of life in rural and remote areas. Aside from mitigating climate change, biomass gasifier systems can contribute to reducing air and water pollution, especially when compared to certain traditional energy sources. This is particularly important for achieving sustainable development objectives related to environmental protection and public health. Biomass gasification presents economic opportunities for communities and regions by promoting the development of local, renewable energy sources and creating jobs and businesses involved in biomass procurement, system setup, and maintenance.

Dynamical model development and parameter identification for the anaerobic digestion (AD) process involve creating a mathematical model that describes the dynamic behaviour of AD and identifying the parameters of the model to fit the data obtained from real or simulated processes. The objective of this process is to develop a model that accurately represents the behaviour of AD, which allows for a better understanding of the underlying biological, chemical, and physical processes. This understanding is crucial for optimising and controlling AD (e.g. when applied to a wastewater treatment process), improving its efficiency, and ensuring the effective removal of organic pollutants and the production of biogas.

The development of a dynamical model typically involves incorporating mass balances, biokinetic equations, and other relevant relationships to describe the dynamics of substrate degradation, microbial growth, biogas production, and other key processes. Once the model is developed, parameter identification is performed to estimate the values of the model parameters by comparing model predictions with experimental or operational data. This step is essential for ensuring that the model accurately reflects the behaviour of the actual system.

Overall, dynamical model development and parameter identification for AD processes are important for advancing process understanding, optimising system performance, and driving improvements in AD technology for biogas production.

This section summarises the complementary results presented in the works of Machado Higuera and Sinitsyn (2015) and Machado Higuera (2015), in which the following system is studied:

$$\frac{dX_1}{dt} = (\mu_1(S_1) - \alpha D) \ X_1$$
$$\frac{dS_1}{dt} = D \left(S_{1\text{in}} - S_1 \right) - k_1 \ \mu_1(S_1) \ X_1,$$

$$(5.11)$$

which is related to the AD model for biogas production.

The following concepts and definitions can be found in the studies of Acosta-Humánez and coworkers (2010, 2006, 2011), although they are quite well known in the literature concerning *differential Galois theory*.

A differential field (K, ∂) is a field K endowed with a derivation ∂. The field of constants of K, denoted by C_K, corresponds to all the elements of K vanishing under the derivation ∂, and it is assumed to be algebraically closed and of characteristic zero.

Consider a differential equation:

$$f(x, y, y', y'', \ldots y^{(n)}) = 0, \tag{5.12}$$

where f is an analytic function in a complex variable x and its coefficients belong to the differential field K. We say that a function θ satisfying equation (5.12) is either:

(1) *algebraic* over K if θ satisfies a polynomial equation whose coefficients belong to K, or
(2) *primitive* over K if $\theta' \in K$, i.e. $\theta = \int \eta$ for some $\eta \in K$, or
(3) *exponential* over K if $\frac{\theta'}{\theta} \in K$, i.e. $\theta = \exp(\eta)$ for some $\eta \in K$.

A solution of equation (5.12) is called *Liouvillian* if there exist differential fields $K = K_0 \subset K_1 \subset \cdots \subset K_n$ such that $K_i = K_{i-1}(\theta)$, $1 \leq i \leq n$, where θ is algebraic, primitive, or exponential over K.

In other words, Liouvillian functions include elementary functions such as algebraic, logarithmic, trigonometric, hyperbolic, and inverses. In the usual terminology of differential equations, by 'analytical' or 'exact' solutions, we means solutions that can be obtained explicitly, including special functions. For instance, there are analytic or exact solutions that are not Liouvillian, such as Airy functions. Therefore, an *error function* is of Liouvillian type. Thus, we consider here only Liouvillian functions as solutions of nonlinear equations. Acosta-Humánez and Suazo (2013) used Liouvillian functions to obtain explicit *propagators*, while in this section we obtain explicit solutions of Abel equations that are Liouvillian functions.

Our results are concerned with Abel equations of the first kind, which is a nonlinear differential equation (NDE) that is cubic in unknown functions and is written in general form as

$$y' = f_0(x) + f_1(x)y + f_2(x)y^2 + f_3(x)y^3. \tag{5.13}$$

The challenge is searching for Liouvillian solutions to an Abel differential equation of the first kind (5.13). The general Liouvillian solution to the Abel equation (5.13) is an open problem. There are not many results on the construction of Liouvillian solutions of the Abel equation (5.13) in evident form. We mention the work presented by Salinas *et al.* (2013), where a new construction method of Liouvillian solutions of (5.13) is suggested. Acosta-Humánez *et al.* (2015) considered Abel equations of the second kind.

The aim of this section is to establish a coupling between the biogas subsystem (5.11), the method to obtain Liouvillian solutions of Abel equations presented by Salinas *et al.* (2013), and the Hamiltonian algebrisation method presented by Acosta-Humánez (2010); Acosta-Humánez (2009) to investigate the existence of Liouvillian solutions for the Abel equation (5.13).

The section is organised as follows. First, following Salinas *et al.* (2013), we obtain an Abel equation from the NDE modelling of upper and lower solutions to biogas production with special coefficients $f_0(x), f_1(x), f_2(x)$, and $f_3(x)$, as proposed by Machado Higuera (2015). In the first step, we consider the nonlinear system of ODEs of biogas production, which includes seven equations, as given by Bernard *et al.* (2001). In the second step, we decompose the original biogas system into subsystems using the method of upper and lower solutions to reduce the first subsystem from two equations into the Abel equation. In the subsequent section, we employ Hamiltonian algebrisation to transform the Abel equation with transcendental coefficients into a new Abel equation with rational coefficients and then compare the numerical approximations of the Abel equation's solutions with the numerical approximations of the algebrisation of the Abel equation's solutions. Finally, the Liouvillian solutions of the algebrised Abel equation are computed.

5.2.1   *Biogas subsystem associated with the Abel equation*

Biogas is the generic name for a combustible gas mixture produced during the decomposition of organic substances due to an anaerobic microbiological process (methane fermentation). Biogas, the end product of AD mainly consisting of CH_4 (60–70%) and CO_2 (30–40%), provides considerable potential as a versatile carrier of renewable energy, largely due to the wide range of substrates that can be used in the AD process (Sialve *et al.*, 2009). AD has been used to convert biomass into biogas microalgae (Sialve *et al.*, 2009; Mariet *et al.*, 2011). It has also been used in wastewater treatment (Mariet *et al.*, 2011; Bernard *et al.*, 2001). Biogas is widely used as a combustible fuel in Germany, Denmark, China, the United States, and other developed countries. It is used for both home consumption and public transportation.

We consider the following nonlinear system of ODEs that models the process of AD (Bernard *et al.*, 2001):

Biomass balance

$$\frac{dX_1}{dt} = (\mu_1(S_1) - \alpha D) \; X_1 \quad \text{(acidogenesis)}, \tag{5.14}$$

$$\frac{dX_2}{dt} = (\mu_2(S_2) - \alpha D) \; X_2 \quad \text{(methanogenesis)}. \tag{5.15}$$

Substrate balance

$$\frac{dS_1}{dt} = D \left(S_{1in} - S_1 \right) - k_1 \; \mu_1(S_1) \; X_1, \tag{5.16}$$

$$\frac{dS_2}{dt} = D \left(S_{2in} - S_2 \right) + k_2 \; \mu_1(S_1) \; X_1 - k_3 \; \mu_2(S_2) \; X_2. \tag{5.17}$$

Alkalinity balance

$$\frac{dA}{dt} = D \left(A_{\text{in}} - A \right). \tag{5.18}$$

Carbon exchange rate

$$\frac{dC}{dt} = D \left(C_{\text{in}} - C \right) + k_4 \; \mu_1(S_1) \; X_1 + k_5 \; \mu_2(S_2) \; X_2 \tag{5.19}$$

$$- K_{L_a}[C + S_2 - A - K_H P_C]. \tag{5.20}$$

The product $K_H \; P_C = B$ determines the concentration of oxygen dissolved in C.

Net rate of methane production

$$\frac{dF_M}{dt} = k_6 \; \mu_2(S_2) \; X_2 \tag{5.21}$$

where F_M is the methane concentration.

The nonlinear kinetic behavior occurs due to the reaction rates, which are given by $\mu_1(S_1) = \mu_{1\max}\frac{S_1}{S_1 + K_{S_1}}$ (monod-type kinetic) and $\mu_2(S_2) = \mu_{2\max}\frac{S_2}{\frac{S_2^2}{K_{I2}} + S_2 + K_{S_2}}$ (Haldane-type kinetic), representing the bacterial growth rates associated with the two bioprocesses.

In this case, the variables are as follows:

S_1 is an organic substrate concentration [g/l];

X_1 is concentration of acetogenic bacteria [g/l];

S_2 is volatile fatty acids concentration [mmol/l];

X_2 is concentration of methanogenic bacteria [g/l];

A is concentration of alkalinity [mmol/l];

C is total inorganic carbon concentration [mmol/l];

F_M is methane concentration [mmol/l d^{-1}].

Here, $\mu_{1\max} > 0$, $\mu_{2\max} > 0$, $K_{S_1} > 0$, $K_{S_2} > 0$, $K_{I2} > 0$, $A_{\text{in}} > 0$, $B > 0$, $C_{\text{in}} > 0$, $K_{L_a} > 0$, $S_{1\text{in}} > 0$, $S_{2\text{in}} > 0$, and $k_i > 0$, $i = 1, 2, 3, 4, 5, 6$, are positive constants. The parameter α $(0 \le \alpha \le 1)$ therefore reflects this process heterogeneity: $\alpha = 0$ corresponds to an ideal fixed-bed reactor, whereas $\alpha = 1$ corresponds to an ideal continuously stirred tank reactor (CSTR) (Benyahia *et al.*, 2012; Bernard *et al.*, 2001).

Definition 5.2.1. A trivial solution of system (5.14)–(5.21) has the form

$$E_1(0, S_{1\text{in}}), \quad E_2(0, S_{1\text{in}}, 0, S_{2\text{in}}), \quad E_3(0, S_{2\text{in}})$$

$$E_4(0, S_{1\text{in}}, 0, S_{2\text{in}}, A_{\text{in}}), \quad E_5(A_{\text{in}}),$$

where $X_1 = 0$, $X_2 = 0$, $S_1 = S_{1\text{in}}$, $S_2 = S_{2\text{in}}$, $A = A_{\text{in}}$, $C = C_{\text{in}}$, and $F_M = 0$.

We are interested in solving the nonlinear subsystem with initial conditions, which models the dynamics of the biomass and the substrate in acidogenesis:

$$\begin{cases} \dfrac{dX_1}{dt} = \left(\mu_{1\max} \dfrac{S_1}{S_1 + K_{S_1}} - \alpha D \right) X_1 \triangleq F\left(t, X_1(t), S_1(t)\right), \\[2mm] \dfrac{dS_1}{dt} = D\left(S_{1\text{in}} - S_1\right) - k_1\, \mu_{1\max} \dfrac{S_1}{S_1 + K_{S_1}} X_1 \triangleq G(t, X_1(t), S_1(t)), \\[2mm] X_1(0) = c_1, \\[2mm] S_1(0) = c_2, \end{cases}$$

$$(5.22)$$

where $t \in [0, T] = I$, with $T > 0$, and F and G are functions of the $C^0(I) = C(I)$ class.

Definition 5.2.2 (lower-upper solution). A pair $[(X_{10}, S_{10}), (X_1^0, S_1^0)]$ is called the lower-upper solution of problem (5.22) if the following conditions are satisfied:

$$(X_{10}, S_{10}) \in C^1(I), \quad (X_1^0, S_1^0) \in C^1(I), \quad t \in I;$$

$$\dot{X}_{10} - F(t, X_{10}, S_1) \leq 0 \quad \text{(lower)};$$

$$\dot{X}_1^0 - F(t, X_1^0, S_1) \geq 0 \text{ in } I, \quad \forall S_1 \in [S_1^0, S_{10}], \quad \text{(upper)};$$

$$\dot{S}_{10} - G(t, X_1, S_{10}) \leq 0 \quad \text{(reverse order)};$$

$$\dot{S}_1^0 - G(t, X_1, S_1^0) \geq 0 \text{ in } I, \forall X_1 \in [X_{10}, X_1^0] \quad \text{(reverse order)}$$

and with initial conditions

$$X_{10}(0) \leq c_2 \leq X_1^0(0), \quad S_1^0(0) \leq c_1 \leq S_{10}(0);$$

with $X_{10} \leq X_1^0$, $S_1^0 \leq S_{10}$ in I.

We reduce system (5.22) via transformation $S_1 = S_{1\text{in}}$ into the nonlinear equation

$$\frac{dX_1}{dt} = \left(\mu_{1\text{max}} \frac{S_1}{S_1 + K_{S_1}} - \alpha D \right) X_1$$

with the solution

$$X_1 = X_1(0) \exp \left(\mu_{1\text{max}} \frac{S_{1\text{in}}}{S_{1\text{in}} + K_{S_1}} - \alpha D \right) t,$$

$$X_1 = X_1(0) \exp (m\, t) \quad \text{with } m = \frac{\mu_{1\text{max}} S_{1\text{in}}}{S_{1\text{in}} + K_{S_1}} - \alpha D.$$

We take $\Gamma > 0$; therefore, the solutions

$$X_1^0 = X_1 + \Gamma \quad \text{and} \quad X_{10} = X_1 - \Gamma$$

represent the upper and lower solutions of (5.22), respectively.

From Definition 5.2.2 to S_1 in (5.22),

$$\dot{S}_{10} - D\left(S_{1\text{lin}} - S_{10}\right) + k_1\,\mu_{1\text{max}}\frac{S_{10}}{S_{10} + K_{S_1}}\,X_1^0 \leq 0 \quad \text{(lower)},$$

$$\dot{S_1}^0 - D\left(S_{1\text{lin}} - S_1^0\right) + k_1\,\mu_{1\text{max}}\frac{S_1^0}{S_1^0 + K_{S_1}}\,X_{10} \geq 0 \quad \text{(upper)}. \tag{5.23}$$

We analyse the case where $X_{10} \neq 0$ on (5.22) to search for an upper solution to S_1. Then,

$$\frac{dS_1^0}{dt} = D(S_{1\text{lin}} - S_1^0) - \left[\frac{k_1\,\mu_{1\text{max}}\,S_1^0}{S_1^0 + K_{S_1}}\right]X_{10},$$

taking X_1^0 from (5.2.1),

$$\frac{dS_1^0}{dt} = D(S_{1\text{lin}} - S_1^0) - \left[\frac{k_1\,\mu_{1\text{max}}\,S_1^0}{S_1^0 + K_{S_1}}\right][X_1(0)\,\exp{(m\,t)} - \Gamma], \tag{5.24}$$

we have $V = (S_1^0 + K_{S_1})^{-1}$ is obtained an Abel equation of the first kind:

$$\frac{dV}{dt} = DV + [-D(S_{\text{in}} + K_{S_1}) + k_1\mu_{1\text{max}}X_1(0)\exp{(mt)} - k_1\mu_{1\text{max}}\Gamma]\,V^2$$

$$+ [-k_1\,\mu_{1\text{max}}\,X_1(0)\,K_{S_1}\exp{(m\,t)} + k_1\mu_{1\text{max}}\Gamma]\,V^3. \tag{5.25}$$

5.2.2   *Algebrisation of Abel equations*

In this section, we transform equation (5.25), an Abel equation, into a new Abel equation with rational coefficients. To do this, we use the *Hamiltonian algebrisation*. For suitability, hereinafter, we write ∂_t instead of $\frac{d}{dt}$.

Definition 5.2.3 (Hamiltonian change of variable). The change of variable $w = w(t)$ is called Hamiltonian if there exists $\alpha = \alpha(w)$ such that $(\partial_t w)^2 = \alpha(w)$. If $\alpha(w) \in \mathbb{C}(w)$, we say that $w = w(t)$ is a rational Hamiltonian change of variable.

The process of obtaining an algebraic form of a differential equation is called the *algebrisation of differential equation*. This means that we transformed a differential equation with non-rational coefficients into a differential equation with rational coefficient. The algebrisation is called Hamiltonian if it is carried out through a Hamiltonian change of variable. Following

Acosta-Humánez *et al.* (2011), we obtain in a natural way the following result.

Proposition 5.2.1. *The algebraic form of the Abel equation* (5.13), *through the Hamiltonian change of variable* $w = \exp(mt)$, *is given by*

$$\partial_w \widehat{V} = \frac{D}{mw} \widehat{V} + \left[\frac{-D(S_{\text{in}} + K_{S_1}) + k_1\mu_{1\text{max}}X_1(0)w - k_1\mu_{1\text{max}}\Gamma}{mw} \right] \widehat{V}^2$$

$$+ \left[\frac{-k_1\,\mu_{1\text{max}}\,X_1(0)\,K_{S_1}w + k_1\mu_{1\text{max}}\Gamma}{mw} \right] \widehat{V}^3, \qquad (5.26)$$

Proof. Due to the change of variable $w = \exp(mt)$ being Hamiltonian, we can obtain $\alpha = \alpha(w) = (\partial_t w)^2$. In this case, $\alpha = m^2 w^2 \in \mathbb{C}(w)$. Furthermore, ∂_t is transformed into $\sqrt{\alpha}\partial_w = mw\partial_w$, which allow us to arrive at our result. That is, from the change of variable

$$w = e^{mt},$$

with

$$\frac{1}{mw}\frac{dV}{dt} = \partial_w \widehat{V},$$

we obtain

$$\partial_w \widehat{V} = \frac{D}{mw} \widehat{V} + \left[\frac{-D(S_{\text{in}} + K_{S_1}) + k_1\mu_{1\text{max}}X_1(0)w - k_1\mu_{1\text{max}}\Gamma}{mw} \right] \widehat{V}^2$$

$$+ \left[\frac{-k_1\,\mu_{1\text{max}}\,X_1(0)\,K_{S_1}w + k_1\mu_{1\text{max}}\Gamma}{mw} \right] \widehat{V}^3. \qquad \square$$

Remark 5.2.2. The algebrised Abel equation (5.26) can lead us to obtaining the solutions for the Abel equation (5.13). Moreover, the Hamiltonian change of variable $w = e^{mt}$ and its derivative form a solution curve of the Hamiltonian $H(w,p) = \frac{p^2}{2} - \frac{m^2 w^2}{2} + const.$

We rewrite (5.26) in the form

$$\partial_w \widehat{V} = f_1 \widehat{V} + f_2 \widehat{V}^2 + f_3 \widehat{V}^3,$$

where

$$f_1 = \frac{D}{mw}, \quad f_2 = \frac{a_5 + a_2 w}{mw}, \quad f_3 = \frac{a_4 w - a_3}{mw},$$

$$a_1 = -D(S_{1\text{in}} + K_{S_1}), \quad a_2 = k_1\mu_{1\text{max}}X_1(0),$$

$$a_3 = -k_1\mu_{1\text{max}}\Gamma, \quad a_4 = -K_{S_1}a_2, \quad a_5 = a_1 + a_3. \qquad (5.27)$$

Numerical approximations.   First, we reduce the Abel equation with the change of variable

$$V = \frac{1}{S_1^0 + K_{S_1}} \tag{5.28}$$

to plot $S_1(t)$.

Then, we perform the algebrisation of the Abel equations with the change of variable

$$\hat{V}(w) = \frac{1}{S_1^0 + K_{S_1}},$$

where

$$w = e^{mt}, \quad w(0) = 1,$$

and

$$t = \frac{1}{m} \ln w,$$

to plot $S_1^0(t)$.

Figure 5.1 shows the graph of a Liouvillian solution for the variable S_1 associated with the algebrised Abel equation. The parameters used for the simulation result shown in Figure 5.1 are listed in Table 5.1.

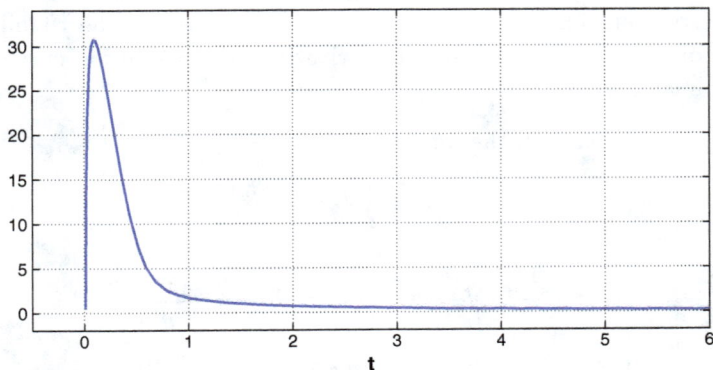

Figure 5.1. S_1^0 via algebrised Abel.

Table 5.1. Parameters for simulation.

Parameter	Value	Units	SD
α	0.5		0.4
D	0.395	$[\text{d}^{-1}]$	0.135
$S_{1\text{in}}$	10.0	$[\text{g/l}]$	6.4
K_{S_1}	12.1	$[\text{g/l}]$	20.62
$\mu_{1\text{max}}$	1.2	$[\text{d}^{-1}]$	
k_1	23.2		

Note: SD = standard deviation.
Source: Bernard *et al.* (2001).

5.2.3 *Exact solutions of Abel equations*

We apply the method presented by Salinas *et al.* (2013) to the construction of Liouvillian solutions for Abel equations in the algebraic form (5.25). The transformation $y = u(x)z(x) + v(x)$ reduces the original Abel equation into the canonical form

$$z' - z^3 - \Phi = 0,$$

with

$$\Phi(\xi) = \frac{\varphi e^{\int \varphi d\xi}}{\sqrt{C - \int e^{2\int \varphi d\xi} d\xi}}$$

$$= \frac{1}{f_3 u^3}\left[f_0 - \frac{f_1 f_2}{3 f_3} + \frac{2 f_2^3}{27 f_3^2} + \frac{1}{3}\frac{d}{dx}\frac{f_2}{f_3}\right],$$

$$u(x) = e^{\int \left(f_1 - \frac{f_2^2}{3 f_3}\right) dx}, \quad \xi = \int f_3 u^2 dx,$$

where φ is an arbitrary function and $v = -\frac{f_2}{3 f_3}$.

Case 1.

$\varphi(\xi) = 0$. In this case, we obtain the solution of the Abel equation (5.26) in the following form:

$$y = \frac{e^{\int \left(f_1 - \frac{f_2^2}{3 f_3}\right) dx}}{\sqrt{C - \int f_3 e^{2\int \left(f_1 - \frac{f_2^2}{3 f_3}\right) dx} dx}} - \frac{f_2}{3 f_3}. \tag{5.29}$$

Case 2.

$\varphi(\xi) = C_0$, where C_0 is a constant.

$$y = e^{\int \left(f_1 - \frac{f_2^2}{3 f_3}\right) dx}\frac{e^{C_0 \int f_3 e^{2\int \left(f_1 - \frac{f_2^2}{3 f_3}\right) dx} dx}}{\sqrt{C - \int e^{2C_0 \int f_3 e^{2\int \left(f_1 - \frac{f_2^2}{3 f_3}\right) dx} dx}}} - \frac{f_2}{3 f_3}. \tag{5.30}$$

To obtain Liouvillian solutions, we consider the phenomenon of vanishing, when the variable $X_1 \longrightarrow 0$. In this case, $X_1(0) = 0$, and we reduce solution (5.29) to the form

$$\hat{V} = \frac{e^{F_1(w)}}{\sqrt{C - F_2(w)}} - \frac{f_2}{3f_3},$$

with

$$F_1(w) = \int \left(f_1 - \frac{f_2^2}{3f_3} \right) dw = \ln \left| C_1 w^{b_1} \right|, \quad F_2 = 2F_1(w),$$

$$F_3(w) = \int f_3 e^{2F_1(w)} dw = \frac{a_3 C_1^2}{2b_1 m} w^{2b_1} + C_2$$

and

$$b_1 = \frac{D}{m} - \frac{a_5^2}{3ma_3}, \quad b_2 = \frac{C_0 a_3 C_1^2}{b_1 m},$$

with constants as defined in (5.27). Applying the change of variable $\hat{V}[e^{mt}] = V(t)$, the Liouvillian solution to case 1 of the Abel equation is

$$V(t) = \frac{C_1 e^{b_1 mt}}{\sqrt{C - 2(\ln C_1 + b_1 mt)}} - \frac{a_5}{3a_3}.$$

Taking up again the change of variable (5.28), the solution to (5.24) is

$$S_1^0 = \frac{3a_3 \sqrt{C - 2(\ln C_1 + b_1 mt)}}{3a_3 C_1 e^{b_1 mt} - a_5 \sqrt{C - 2(\ln C_1 + b_1 mt)}} - K_{S-1}.$$

Figure 5.2 shows the graph of a Liouvillian solution for the variable S_1 associated with case 1 of the Abel equation.

Now, reduce solution (5.30) to the form

$$\hat{V} = e^{F_1(w)} \frac{e^{c_0} F_3(w)}{\sqrt{C - F_4(w)}} - \frac{f_2}{3f_3},$$

with

$$F_4(w) = \int e^{2C_0 F_3(w)} dw = -\frac{e^{2C_0 C_2}}{2b_1(-b_2)^{1/2b_1}} \Gamma \left(\frac{1}{2b_1}, -b_2 w^{2b_1} \right) + C_3.$$

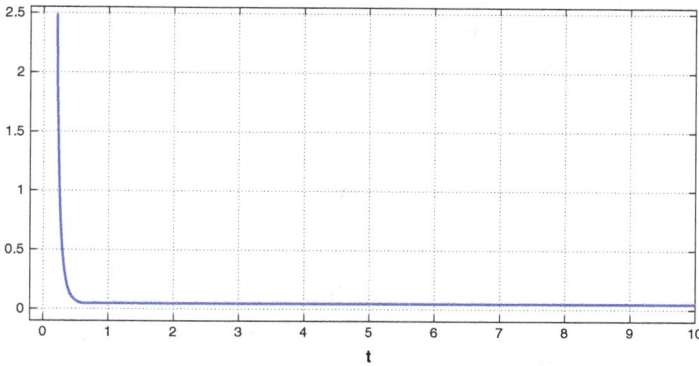

Figure 5.2. S_1^0 via Abel equation, case 1.

Using the change of variable above, the solution to case 2 of the Abel equation is

$$V(t) = C_1 e^{b_1 mt} \frac{\frac{C_0 a_3 C_1^2}{2b_1 m} e^{2b_1 mt} + C_2}{\sqrt{C - C_3 + \frac{e^{2C_0 C_2}}{2b_1 (-b_2)^{1/2b_1}} \Gamma\left(\frac{1}{2b_1}, -b_2 e^{2mb_1}\right)}} - \frac{a_5}{3a_3}.$$

From (5.28), the solution to S_1^0 is

$$S_1^0 = \frac{3a_3 \sqrt{C - C_3 + \frac{e^{2C_0 C_2}}{2b_1 (-b_2)^{1/2b_1}} \Gamma\left(\frac{1}{2b_1}, -b_2 e^{2b_1 mt}\right)}}{3a_3 C_1 e^{b_1 mt} \left(\frac{C_0 a_3 C_1^2}{2b_1 m} e^{2b_1 mt} + C_2\right)} \\ - a_5 \sqrt{C - C_3 + \frac{e^{2C_0 C_2}}{2b_1 (-b_2)^{1/2b_1}} \Gamma\left(\frac{1}{2b_1}, -b_2 e^{2b_1 mt}\right)}.$$

In this section, we found a direct connection between the variable that represents the substrate in the acidogenesis process and the Abel equation of the first kind. In the search for Liouvillian solutions to this equation, we implemented a combination of Hamiltonian algebrisation and a method to obtain Liouvillian solutions of the Abel equations given by Salinas *et al.* (2013). We found Liouvillian solutions for cases 1 and 2. The figures show that the solution of the differential equation for S_1 has the same behaviour as the numerical approximations, which tells us that these functions represent the process.

5.3 Singular Integral Equations of Abel Type: Numerical Solution and Application to Infrared Flame Tomography

Flame tomography, also known as flame imaging or flame spectroscopy, is important in combustion theory and combined heat and power (CHP). Let us outline its main fields of application:

(1) *Understanding combustion processes*: Flame tomography allows researchers and engineers to visualise and study the complex processes occurring in combustion, such as ignition, flame stabilisation, and pollutant formation. This understanding is crucial for optimising combustion efficiency, reducing emissions, and developing cleaner and more efficient energy production technologies.
(2) *Emission control*: By providing detailed information about the flame structure and combustion dynamics, flame tomography helps in designing more effective emission control strategies. This is crucial for ensuring compliance with environmental regulations and minimising the environmental impact of combustion processes.
(3) *CHP system optimisation*: In CHP plants, efficient combustion is crucial for maximising the utilisation of fuel and minimising waste heat. Flame tomography can provide insights into flame behaviour, temperature distribution, and combustion stability, which are all essential for optimising CHP system performance.
(4) *Safety and reliability*: Understanding the behaviour of flames in combustion processes is essential for ensuring the safety and reliability of combustion systems. Flame tomography can help in identifying potential issues such as flame instability, flashback, or incomplete combustion, allowing for the design of safer and more reliable combustion systems.

Overall, flame tomography is an important tool in combustion theory and CHP design, providing valuable insights into combustion processes and helping to improve the efficiency, environmental performance, and safety of combustion systems. This section focuses on one of the principal mathematical models employed in flame tomography described by a special class of singular integral equations (SIEs) of the first kind.

Numerical methods for solving a variety of SIEs are offered in many publications. A one-dimensional SIE of the first and second kinds with Cauchy kernels, Hilbert kernels, logarithmic, etc. as well as two-dimensional, nonlinear SIEs, have been addressed. Here, we concentrate on the Abel SIE

$$2 \int_x^R \frac{r}{\sqrt{r^2 - x^2}} k(r) \, dr = q(x), \quad 0 \le x \le R, \tag{5.31}$$

where $k(r)$ is the desired function and $q(x)$ is the source function. SIEs in the form of equation (5.31) are widely used in practical models, including in plasma diagnostics, thermal tomography, X-ray CT, spectroscopy, and galaxy cluster astrophysics. In all these problems, the object of interest enjoys axial (or spherical) symmetry. The Abel equation can also be written as

$$\int_0^x \frac{k(r)}{\sqrt{x - r}} \, dr = q(x), \quad 0 \le x \le R. \tag{5.32}$$

Equation (5.32) describes various problems in mechanics (such as the tautohron problem), scattering, and other problems. Of course, one may transfer SIE (5.31) into SIE (5.32) and vice versa, but it makes it more complicated to analyse their physical meaning.

In the following, we outline the main algorithms for the numerical solution of SIEs and singular integrals computation:

(1) *Algorithms based on relevant mesh shift*: Belotserkovskii and Lifanov (1993) and Boikov and Kudryashova (2000) introduced the discrete meshes of knots with respect to variables r and x, i.e. $r_j = jh$, $x_i = r_i + \Delta$, $j, i = 0, 1, \ldots, n$, $r_n = R$, where the step $h = R/n$ and Δ is a mesh shift, which is $h/2$ (Belotserkovskii and Lifanov, 1993) or $\Delta \in (0, h/2)$ (Boikov and Kudryashova, 2000). The introduction of the shift Δ enables us to overcome singularity when it comes to applying quadrature rules. However, such algorithms need this shift selection.

(2) *Quadrature type methods*: One of the popular methods is the discrete vortices method (for more details, readers may refer to the work of Belotserkovskii and Lifanov, 1993), where an integral with the Cauchy kernel

$$\frac{1}{2\pi} \int_{-1}^1 \frac{\gamma(x)}{x - x_0} \, dx = f(x_0), \quad -1 < x_0 < 1,$$

is approximated with lift rectangles quadrature rule and using meshes on x and x_0 with a shift of $\Delta = h/2$. This gives a system of linear algebraic equations (SLAE) with a non-zero main diagonal.

Daun *et al.* (2006) suggested the 'onion peeling' method for solving SIEs (5.31). In this method, the region $r \in [0, R]$ is approximated with rings Δr wide of constant values $k \in (r_j - \Delta r/2, r_j + \Delta r/2)$ for each r_j. Here, meshes are assumed to be uniform ($\Delta = 0$). The main idea in this method is that the integral $\int_{r_j - \Delta r/2}^{r_j + \Delta r/2} \frac{r}{\sqrt{r^2 - x_i^2}} dr, r_j - \Delta r/2 \geq x_i$ is computed analytically and is finite. Further, a midpoint quadrature is used, resulting in systems of linear algebraic equations with an upper triangular matrix with respect to $k_j = k(x_j)$. A similar method is suggested by Sizikov *et al.* (2004).

(3) *Solution approximation*: Anderssen and de Hoog (1990), Minerbo and Levy (1969), Preobrazhensky and Pikalov (1982), Saadamandi and Dehghan (2008), Singh *et al.* (2009), and Voskoboynikov *et al.* (1984), among others, approximated the desired solution $k(r)$ (as well as the right-hand side $q(x)$) with an orthogonal polynomial, shifted Legendre polynomials, normalised Bernstein polynomials, algebraic or trigonometric polynomial, or polynomial spline consisting of coefficients determined with minimum discrepancy between the left- and right-hand sides of (5.31). This leads to projection methods (such as the Galerkin method, the collocation method, the method of splines, the quadrature method, and the least squares method) and to the solution of a SLAE wrt the corresponding polynomial coefficients.

These algorithms feature *self-regularisation*, and in the case of using the relative shift of meshes, the shift Δ plays the role of a regularisation parameter. Namely, if Δ is closer to $h/2$, then the solution $k(r)$ is more stable, but it reduces thef resolving capability of the method. If Δ is closer to zero, then the solution is less stable, but the resolving capability of the method is higher. In all these algorithms, a SLAE contains a prevailing (but not infinite) matrix diagonal.

Equation (5.31), as is known, has an analytical solution:

$$k(r) = -\frac{1}{\pi} \int_r^R \frac{q'(x)}{\sqrt{x^2 - r^2}} dx, \quad 0 \leq r \leq R. \tag{5.33}$$

However, solution (5.33) contains the derivative $q'(x)$ of the experimental (noisy) function $q(x)$ and the problem of differentiation is ill posed (see also Chapter 1 of this book). Moreover, the integral in (5.33) is improper (singular). Nevertheless, a number of algorithms, as listed in the following, have been proposed to compute the solution according to (5.33).

(4) *Interpolation and quadrature method*: Nekrasov (1951) computed the derivative $q'(x)$ using interpolation on three (and two) neighboring points (i.e. discrete values of x). The integral $\int_r^R \frac{dx}{\sqrt{x^2-r^2}}$ (cf. (5.33)) is computed analytically (without singularity). A similar algorithm was suggested by Sizikov *et al.* (2004) using the generalised left rectangles formula.

(5) *Smoothing polynomials method*: The function $q(x)$ can be approximated by a linear combination of smoothing polynomials (or splines) uniform over the whole interval $x \in [0, R]$. The derivative $q'(x)$ is computed using polynomial (or spline) differentiation. The solution $k(r)$, in accordance with (5.33), is computed by summing the integral in (5.33) along segments that performed analytically.

(6) *Algorithm without using derivative $q'(x)$*: In the work of Deutsch and Beniaminy (1982), formula (5.33) was converted (by means of integration by parts) into the following expression that does not contain the derivative $q'(x)$:

$$k(r) = -\frac{1}{\pi} \left\{ \frac{q(R) - q(r)}{\sqrt{R^2 - r^2}} + \int_r^R \frac{x\,[q(x) - q(r)]}{\sqrt{(x^2 - r^2)^3}}\, dx \right\}, \quad 0 \le r \le R.$$

(7) *Use of regularisation*: The Abel equation (5.31) enjoys the self-regularising property due to singularity; as a result, the problem of its solving is moderately ill posed (Daun *et al.*, 2006). This means that the above-mentioned algorithms are moderately stable.

In this work, we develop the following variant to numerically solving certain SIEs. We make the meshes of nodes in r and x coincide (i.e. $\Delta = 0$) and eliminate the singularities using the generalised quadrature formula. However, such a technique cannot be applied to all SIEs. For example, it is not applicable to SIEs with the Cauchy kernel, whereas it is applicable to some SIEs with logarithmic and other (weakly singular) kernels. section function $k(r)$, as well as numerical computation of $k(r)$ according to (5.33) using the generalised quadrature method through Tikhonov regularisation.

5.3.1 The generalised quadrature method

We describe the method of using generalised left rectangles formula applied to numerically solve SIE (5.31) (the first method) and to compute $k(r)$ according to (5.33) (the second method).

In this section, we employ nonuniform meshes and left rectangles, resulting in a more generic and convenient algorithm. The solution error estimates for equation (5.31) using the generalised quadrature method are also derived. This method is described as follows in two variants (the first and second methods).

First quadrature method: The first quadrature method employs the generalised left rectangle formula. Let us introduce nonuniform (but coinciding) meshes on x and r as follows:

$$0 = x_1 = r_1 < x_2 = r_2 < \cdots < x_i = r_i < \cdots < x_n = r_n = R. \qquad (5.34)$$

Here, $R = r_{\max}$ is a boundary value such that $k(R + 0) = 0$. On each interval $[r_j, r_{j+1})$, $j = 1, 2, \ldots, n - 1$, we suppose that, approximately,

$$k(r) = k(r_j) \equiv k_j = \text{const.} \qquad (5.35)$$

We have the following.

Lemma 5.3.1. *Under condition (5.35), one has the equality*

$$\int_{r_j}^{r_{j+1}} \frac{r}{\sqrt{r^2 - x^2}} \, k(r) \, dr = \left(\sqrt{r_{j+1}^2 - x^2} - \sqrt{r_j^2 - x^2} \right) k_j,$$

$$j \in [1, n - 1], \quad x \le r_j < r_{j+1} \le R. \qquad (5.36)$$

Proof. The proof is obvious and based on the table integral

$$\int \frac{r}{\sqrt{r^2 - x^2}} \, dr = \sqrt{r^2 - x^2} \quad \text{for } x \le r. \qquad \square$$

Definition 5.3.1. We call formula (5.36) the *generalised quadrature formula of left rectangles* (cf. Krylov, 2005) for the specific singularity $r/\sqrt{r^2 - x^2}$, and the multipliers $\sqrt{r_{j+1}^2 - x^2} - \sqrt{r_j^2 - x^2}$ are the *quadrature coefficients* of this singularity.

The specifics of formula (5.36) are that the singular integral $\int_{r_j}^{r_{j+1}} \frac{r}{\sqrt{r^2 - x^2}} \, dr$ is accurately calculated analytically without peculiarity. If it is calculated numerically using the usual left rectangles quadrature formula, then at $x = r_j$, there will be a division by zero.

Let us now formulate the main result as follows.

Theorem 5.3.2. *A numerical solution for equation* (5.31) *according to the first quadrature method is defined as the following recursion:*

$$
\begin{cases}
k_{n-1} = \dfrac{q_{n-1}/2}{p_{n-1,n-1}}, \\[2ex]
k_i = \dfrac{q_i/2 - \displaystyle\sum_{j=i+1}^{n-1} p_{ij}k_j}{p_{ij}}, & i = n-2,\ n-3,\ldots,1, \\[3ex]
k_n = k_{n-2} + \left(\dfrac{r_n - r_{n-2}}{r_{n-1} - r_{n-2}}\right)(k_{n-1} - k_{n-2}),
\end{cases}
\tag{5.37}
$$

where $k_i \equiv k(r_i)$, $q_i \equiv q(x_i)$,

$$
p_{ij} = \sqrt{r_{j+1}^2 - x_i^2} - \sqrt{r_j^2 - x_i^2},
\tag{5.38}
$$

Proof. The integral in (5.31) is the sum of integrals (5.36), i.e.

$$
\int_{x_i}^{R} \frac{r}{\sqrt{r^2 - x_i^2}}\, k(r)\, dr = \sum_{j=i}^{n-1} \left(\sqrt{r_{j+1}^2 - x_i^2} - \sqrt{r_j^2 - x_i^2}\right) k_j = q_i/2,
\tag{5.39}
$$

$$
i = 1, 2, \ldots, n-1.
$$

This is the SLAE wrt $\{k_j\}_{j=1}^{n}$. The SLAE (5.39) is upper triangular, and its solution can be recursively constructed. From (5.39), for $i = n-1$, $n-2, \ldots, 1$, we eventually obtain $k_{n-1}, k_{n-2}, \ldots, k_1$ according to (5.37). As for the value of $k_n \equiv k(R)$, it cannot be found using this scheme; however, it can be additionally determined as $k_n = 0$ from physical concepts, $k_n = k_{n-1}$, or can be derived using linear extrapolation, as was done in (5.37). □

Daun *et al.* (2006) presented formulas of type (5.36) but for the case of uniform (and coinciding) meshes in r and x and using the middle rectangles quadrature formula. Furthermore, important formulas of type (5.37) were not given.

We call the method involving (5.34)–(5.39) for solving equation (5.31) the *generalised quadrature method* for solving SIE (5.31).

Second quadrature method. Let us here consider the second method, which is the generalised quadrature method designed to compute the singular integral (5.33) giving the solution $k(r)$. It is assumed that $k'(x)$ is computed using some stable method. As in the first method, we introduce node meshes (5.34). On each $[x_i, x_{i+1})$, $i = 1, 2, \ldots, n-1$, it is assumed that

$$q'(x) = q'(x_i) \equiv q_i' = \text{const}. \tag{5.40}$$

The following Lemma can be formulated.

Lemma 5.3.3. *Under condition* (5.40), *the following equality is true:*

$$\int_{x_i}^{x_{i+1}} \frac{q'(x)}{\sqrt{x^2 - r^2}} \, dx = \ln \frac{x_{i+1} + \sqrt{x_{i+1}^2 - r^2}}{x_i + \sqrt{x_i^2 - r^2}} \, q_i', \tag{5.41}$$

$$i = 1, 2, \ldots, n-1, \quad r \le x_i < x_{i+1} \le R.$$

Proof. The integral $\int \frac{dx}{\sqrt{x^2 - r^2}} = \ln\left(x + \sqrt{x^2 - r^2}\right)$ for $r \le x$ is the table integral. Taking into account condition (5.40), we obtain (5.41). $\square$

Definition 5.3.2. Formula (5.41) is the generalised left rectangle quadrature rule (GLRQR) for the singularity $1/\sqrt{r^2 - x^2}$, and the multipliers $\ln \frac{x_{i+1} + \sqrt{x_{i+1}^2 - r^2}}{x_i + \sqrt{x_i^2 - r^2}}$ are the *quadrature coefficients* of this singularity.

Now, we can formulate the following theorem.

Theorem 5.3.4. *A numerical solution of SIE* (5.31) *using formula* (5.33) *according to the second generalised quadrature method results from the following recurrence formulas:*

$$\begin{cases} k_j = -\frac{1}{\pi} \sum_{i=j}^{n-1} g_{ij} \, q_i', \quad j = 2, 3, \ldots, n-1, \\ k_1 = \dfrac{q_1/2 - \sum_{j=2}^{n-1}(r_{j+1} - r_j) k_j}{r_2}, \end{cases} \tag{5.42}$$

where

$$g_{ij} = \ln \frac{x_{i+1} + \sqrt{x_{i+1}^2 - r_j^2}}{x_i + \sqrt{x_i^2 - r_j^2}}. \tag{5.43}$$

Proof. The integral in (5.33) is the sum of integrals (5.41) over the separate intervals $[x_i, x_{i+1})$, i.e.

$$\int_{r_j}^{R} \frac{q'(x)}{\sqrt{x^2 - r_j^2}} \, dx = \sum_{i=j}^{n-2} \ln \frac{x_{i+1} + \sqrt{x_{i+1}^2 - r_j^2}}{x_i + \sqrt{x_i^2 - r_j^2}} \, q_i', \quad j = 1, 2, \ldots, n-1.$$

(5.44)

As a result, solution (5.33) in the discrete form is $\{k_j\}_{j=2}^{n-1}$ according to (5.42) and (5.43). For $j = 1$, this (second) method due to (5.44) gives an uncertainty of $\infty \cdot 0$ for $i = j = 1$ since $x_1 = r_1 = q_1' = 0$. In this case, let's use the first method to determine k_1. Using (5.37) and (5.38) for $i = 1$, we find k_1, in view of (5.42). As for k_n, it can be calculated through linear extrapolation (see (5.37)). □

The advantage of the above two methods is that they do not require the relative shift of meshes, and their integrals with singularities $r/\sqrt{r^2 - x^2}$ and $1/\sqrt{r^2 - x^2}$ are calculated analytically and without divergences.

However, these methods are not suitable for all singularities, such as for numerical computation of hypersingular integrals with the Cauchy kernel $\int_{-1}^{1} \frac{x(\tau)}{\tau - t} \, d\tau$.

5.3.2 *Error estimates*

Let us derive errors estimates for the solution of SIE (5.31).

Quadrature error in a small interval. The quadrature error for computing integral (5.36) on a separate small interval, $[r_j, r_{j+1})$, can be estimated by taking into account the approximation (5.35). Here, no measurement errors for $q(x)$ are assumed.

Lemma 5.3.5. *The integral*

$$\int_{r_j}^{r_{j+1}} \frac{r}{\sqrt{r^2 - x_i^2}} k(r) \, dr, \quad x_i \leq r_j < r_{j+1} \leq R,$$

(5.45)

$$i = 1, 2, \ldots, n-1, \quad j = i, \ldots, n-1,$$

when using the generalised left rectangles formula (5.36) and taking account of the quadrature error caused by the approximation (5.35) is equal to (a refinement of formula (5.36))

$$\int_{r_j}^{r_{j+1}} \frac{r}{\sqrt{r^2 - x_i^2}} k(r) \, dr = p_{ij} k_j + \Delta \varepsilon_{ij},$$

(5.46)

where p_{ij} are quadrature coefficients (5.38) and $\Delta\varepsilon_{ij}$ is the quadrature error in the computation of integral (5.45) and is approximately equal to

$$
\Delta\varepsilon_{ij} = \frac{k'(\xi_j)}{2}\left[(r_{j+1} - 2r_j)\sqrt{r_{j+1}^2 - x_i^2} + r_j\sqrt{r_j^2 - x_i^2}\right.
$$
$$
\left. + x_i^2 \ln \frac{r_{j+1} + \sqrt{r_{j+1}^2 - x_i^2}}{r_j + \sqrt{r_j^2 - x_i^2}}\right], \quad \xi_j \in [r_j, r_{j+1}). \tag{5.47}
$$

Proof. Using the first method, we assume that $\widetilde{k}(r) = k_j$ $r \in [r_j, r_{j+1})$ (see (5.35), i.e. we represent function $k(r)$ by the interpolation Lagrange zero degree polynomial (Korn and Korn, 1961). The error due to such an interpolation is $\Delta k_j(r) \equiv k(r) - k_j = k'(\xi)(r - r_j)$, where $\xi = \xi_j(r) \in [r_j, r_{j+1})$. Then,

$$
k(r) = k_j + k'(\xi)(r - r_j). \tag{5.48}
$$

Let us now substitute (5.48) into (5.45). We get (5.46), where

$$
\Delta\varepsilon_{ij} = k'(\xi_j) \int_{r_j}^{r_{j+1}} \frac{r(r - r_j)}{\sqrt{r^2 - x_i^2}} \, dr. \tag{5.49}
$$

The integral in (5.49) can be analytically computed, giving us the estimate (5.47). □

It should be noted that the derivative $k'(\xi_j)$ in (5.47) can be approximated with

$$
k'(\xi_j) = \frac{k_{j+1} - k_j}{r_{j+1} - r_j}, \quad j = 1, 2, \ldots, n-1. \tag{5.50}
$$

Quadrature error. Let us estimate the quadrature error of the solution of equation (5.31) due to the approach (5.35) (without the error $q(x)$). We formulate this as a theorem.

Theorem 5.3.6. *The errors in the numerical solution of equation (5.31) using the first generalised quadrature method according to (5.37) are computed with the following recurrence:*

$$
\begin{cases}
\Delta k_n = \Delta k_{n-1} = \dfrac{\Delta\varepsilon_{n-1,n-1}}{p_{n-1,n-1}}, \\[4mm]
\Delta k_i = \dfrac{\varepsilon_i - \sum_{j=i+1}^{n-1} p_{ij}\,\Delta k_j}{p_{ii}}, \quad i = n-2,\, n-3, \ldots, 1,
\end{cases} \tag{5.51}
$$

where

$$\varepsilon_i = \sum_{j=i}^{n-1} \Delta\varepsilon_{ij}. \tag{5.52}$$

Here, p_{ij}, $\Delta\varepsilon_{ij}$, and $k'(\xi_j)$ are computed based on (5.38), (5.47), and (5.50), respectively.

Proof. Let us write the integral in (5.31) as a sum of integrals wrt intervals:

$$\int_{x_i}^{R} \frac{r}{\sqrt{r^2 - x_i^2}} k(r)\, dr = \sum_{j=i}^{n-1} \int_{r_j}^{r_{j+1}} \frac{r}{\sqrt{r^2 - x_i^2}} \left(k_j + \Delta k_j(r)\right) dr$$

$$= \sum_{j=i}^{n-1} \int_{r_j}^{r_{j+1}} \frac{r}{\sqrt{r^2 - x_i^2}} \left(k_j + k'(\xi)\left(r - r_j\right)\right) dr, \quad i = 1, 2, \ldots, n-1.$$

Then,

$$\sum_{j=i}^{n-1} \int_{r_j}^{r_{j+1}} \frac{r}{\sqrt{r^2 - x_i^2}} \Delta k_j(r)\, dr = \sum_{j=i}^{n-1} \int_{r_j}^{r_{j+1}} \frac{r\left(r - r_j\right)}{\sqrt{r^2 - x_i^2}} k'(\xi)\, dr,$$

$$i = 1, 2, \ldots, n-1. \tag{5.53}$$

In order to compute the integral on the left-hand side of (5.53), we employ the formula of left rectangles, i.e. we assume that $\Delta k_j(r) = \Delta k(r_j) \equiv \Delta k_j = \text{const}$, $r \in [r_j, r_{j+1})$, and compute the integral using the generalised formula in a similar way as we did for (5.36). The integral on the right-hand side of (5.53) is equal to $\Delta\varepsilon_{ij}$ due to (5.49). Then,

$$\sum_{j=i}^{n-1} p_{ij}\, \Delta k_j = \varepsilon_i, \quad i = 1, 2, \ldots, n-1, \tag{5.54}$$

where ε_i denotes sum (5.52). Here, (5.54) is a SLAE wrt $\{\Delta k_j\}_{j=1}^{n-1}$. It is also assumed that $\{k_i\}_{i=1}^{n}$ are computed using (5.37) in advance. Then, the values $\{\varepsilon_i\}_{i=1}^{n-1}$ are computed using (5.52), (5.47), and (5.50). We solve the SLAE (5.54) using the upper-triangular matrix and obtain solution (5.51) in the recurrent form, adding the condition $\Delta k_n = \Delta k_{n-1}$. □

It should be noted here that formulas (5.51) give the errors of the solution Δk_i with their signs, in contrast to other works where only the absolute values $|\Delta k_i|$, the upper bounds $|\Delta k_i| \leq \cdots$, or upper bounds by the norm $\|\Delta k_i\| \leq \cdots$, etc. are given. This enables us to obtain the refined solution

$$\widehat{k}_i = k_i + \Delta k_i, \quad i = 1, 2, \ldots, n, \tag{5.55}$$

using $\{k_i\}_{i=1}^n$ from (5.37) and the errors $\{\Delta k_i\}_{i=1}^n$ from (5.51).

The errors in the numerical solution given in (5.51) are obtained with regard to only the quadrature errors and the error on the right-hand side $q(x)$ of equation (5.31) is set as zero. Let us take into account the measurement errors $\{\delta_i\}_{i=1}^n$ of the source function $q(x)$. Krylov *et al.* (1977) and Verlan' and Sizikov (1986) estimated the error in the numerical solution of the Volterra integral equations of the first and second kinds by taking into account both the quadrature and source function errors. In a similar way, we can generalise recurrence formulas (5.51) to the case of the errors $\{\delta_i\}_{i=1}^n$ as follows:

$$\begin{cases} |\Delta k_n| = |\Delta k_{n-1}| = \dfrac{|\Delta \varepsilon_{n-1,n-1}| + \delta_{n-1}}{p_{n-1,n-1}}, \\[4mm] |\Delta k_i| = \dfrac{|\varepsilon_i| + \delta_i + \sum_{j=i+1}^{n-1} p_{ij} |\Delta k_j|}{p_{ii}}, \quad i = n-2, n-3, \ldots, 1. \end{cases} \tag{5.56}$$

However, estimates (5.56) give overstated estimates of $\{|\Delta k_i|\}_{i=1}^n$ due to the use of the operation $|\cdot|$ (absolute value).

5.3.3   *Numerical examples*

In order the validate these two generalised quadrature methods, they have been implemented in MATLAB. Following the first method, we search $\{k_i\}_{i=1}^n$ using (5.37) and (5.38). The errors $\{\Delta k_i\}_{i=1}^n$ are calculated using (5.51), as well as (5.38), (5.47), (5.50), and (5.52). The refined solution $\{\widehat{k}_i\}_{i=1}^n$ is calculated with (5.55). The Tikhonov regularisation

$$k_\alpha = (\alpha E + A^T A)^{-1} A^T f \tag{5.57}$$

is employed, where the discrepancy principle (Morozov, 1984) is used to choose the regularisation parameter $\alpha > 0$:

$$\|A k_\alpha - f\| = \delta. \tag{5.58}$$

Here, $f = q/2$, E is an identity matrix, and A is the matrix of the SLAE (5.39), represented as

$$Ak = f, \tag{5.59}$$

where

$$A_{ij} = \begin{cases} p_{ij}, & j \geq i, \\ 0, & \text{otherwise}, \end{cases} \qquad i,j = 1,2,\ldots,n-1. \qquad (5.60)$$

Following the second method, the solution $k(r)$ has been computed using the singular integral in (5.33) based on the generalised left rectangles formula according to (5.42) and (5.43).

Tomography example. The first method was applied to axially symmetric flame diagnostics using infrared tomography. Figure 5.3 shows the measured output intensity $I_m(x)$ of rays (where m is from measurement),

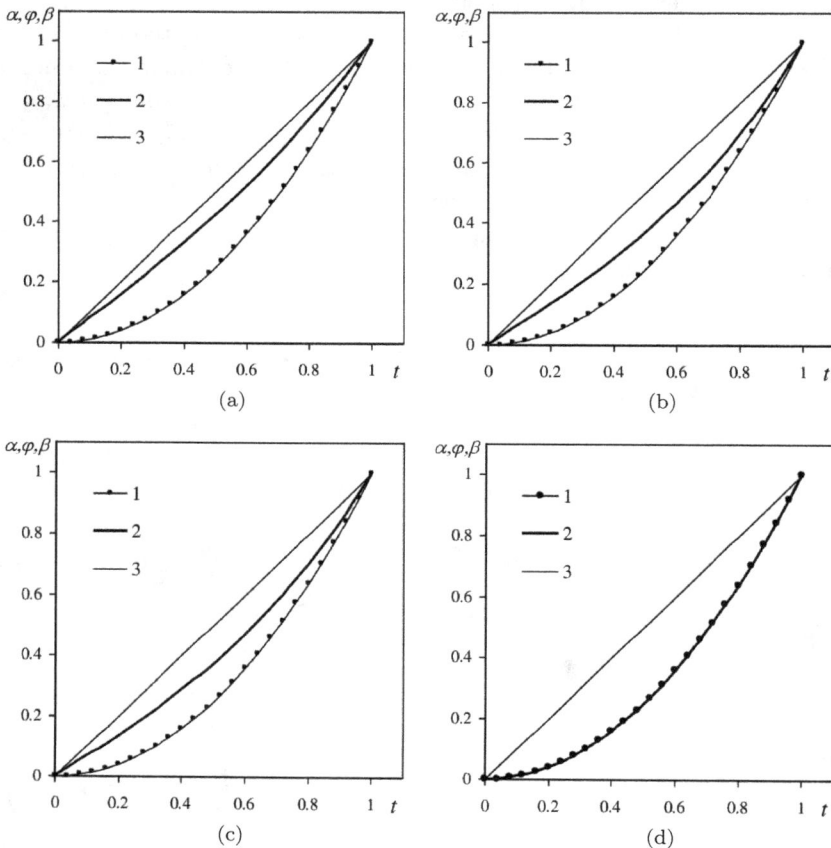

Figure 5.3. Measured noise intensity $I_m(x)$ (the difference between intensity in active and passive regimes). The mesh is nonuniform, with the number of nodes $n = 11$.

which go through gas, undergo absorption and emission, and are intercepted by detectors. The measurements were performed at the Department of Chemical and Biochemical Engineering at the Technical University of Denmark (Risø DTU; before 1 January 2012) as part of a joint project (Evseev *et al.*, 2011, 2013).

The intensity $I_m(x)$ is recalculated as $q_m(x) = -\ln[I_m(x)/B(T_0)]$ (the right-hand side of equation (5.31)), where $B(T_0)$ is the Planck function of the rays source with its temperature $T_0 = 894.4°C$. Figure 5.4 is a plot of the function $q_m(x)$.

Figure 5.5 shows the results of solving SIE (5.31) wrt the absorption coefficient $k_m(r)$ using the first method of generalised quadratures according to (5.37) and (5.38).

As seen, the solution $k_m(r)$ suffers from significant artificial perturbations. This is due to the step of the node mesh being too large (or, in other words, n is too small), as well as measurement errors in the function $I_m(x)$. Here, we also demonstrate the behaviour of the solution $k_{ma}(r)$ derived through Tikhonov regularisation using (5.57), (5.59), (5.60), and (5.39). The regularisation parameter α is chosen using the discrepancy principle (5.58), where $\delta = 0.037$, as a result of which $\alpha = 10^{-0.09} = 0.813$. Figure 5.5 demonstrates that the solution has been smoothed by using the regularisation method.

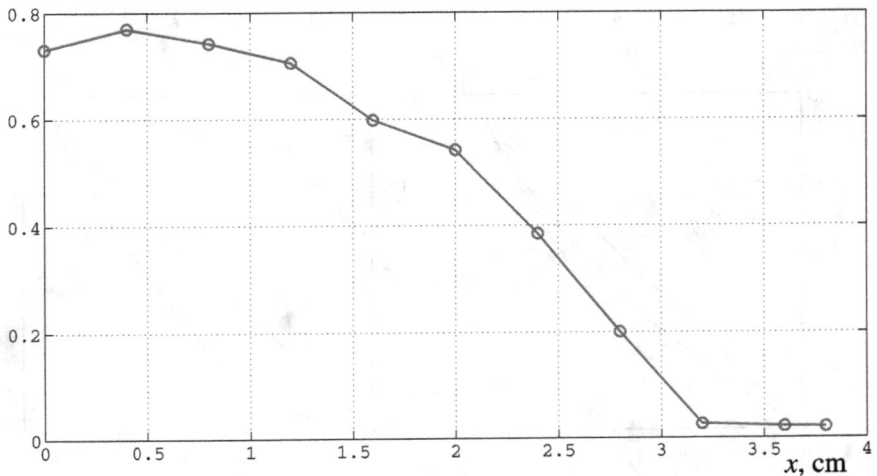

Figure 5.4.   Dimensionless right-hand side $q_m(x)$ of SIE (5.31), with $n = 11$.

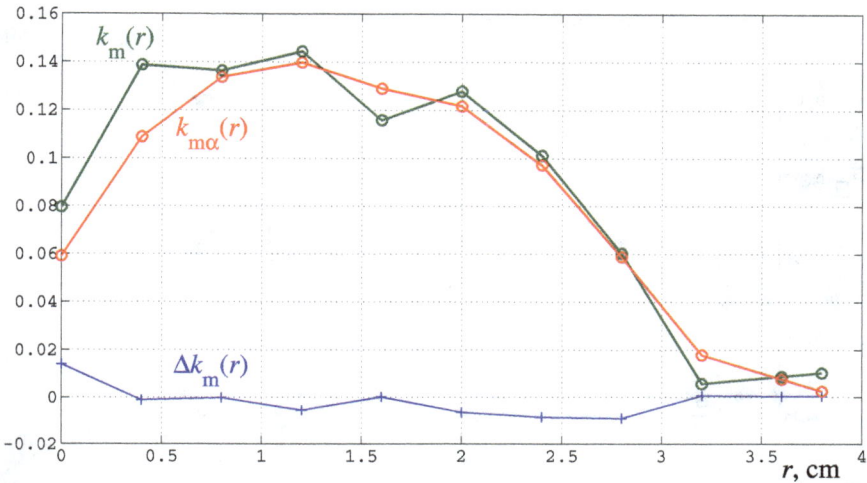

Figure 5.5.   Absorption coefficient $k_m(r)$ computed using the first generalised quadrature method, $k_{m\alpha}(r)$, through Tikhonov regularisation and the errors $\Delta k_m(r)$ of the solution $k_m(r)$ according to (5.51), cm^{-1}.

To reduce the grid step size in x as well as to moderately smooth the fluctuations in the function $I_m(x)$, a spline approximation was used. Figure 5.6 shows an approximation of the function $I_m(x)$ achieved using a cubic smoothing spline.

The smoothed values of $I(x)$ were generated with splines (Fig. 5.6), and then SIE (5.31) was resolved with the generalised quadrature (5.37). Figure 5.7 shows the obtained solution $k(r)$. We also obtained the solution using Tikhonov regularisation (5.57) for $\alpha = 10^{-2}$. Figure 5.7 shows the regularised solution $k_\alpha(r)$.

Figures 5.6 and 5.7 demonstrate that the application of spline-based smoothing enables us to reduce the mesh step size for x (causing an increase in n). This allows for (moderate) smoothing of $k(r)$ and $k_\alpha(r)$. Moreover, in the case of noisy $I(x)$ and large step sizes of the mesh, regularisation slightly improves the solution, as shown in Figure 5.5. In the case of <1% errors and small step sizes (Fig. 5.6), the solutions $k(r)$ (without regularisation) and $k_\alpha(r)$ (with regularisation) are obtained to be practically identical (Fig. 5.7). It confirms that the problem of solving the SIEs is moderately ill posed and has the property of self-regularisation.

In this section, we outlined two new methods of numerically solving SIEs of the Abel type. The methods are based on the use of generalised

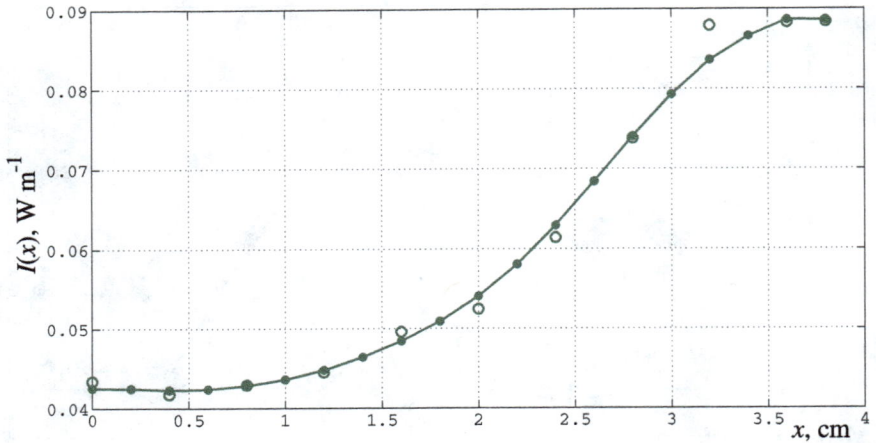

Figure 5.6.   Measured $I_m(x)$ values are marked with ○ ($n = 11$), $I_m(x)$ values for $n = 20$ marked with •, and spline interpolated values are marked with solid line.

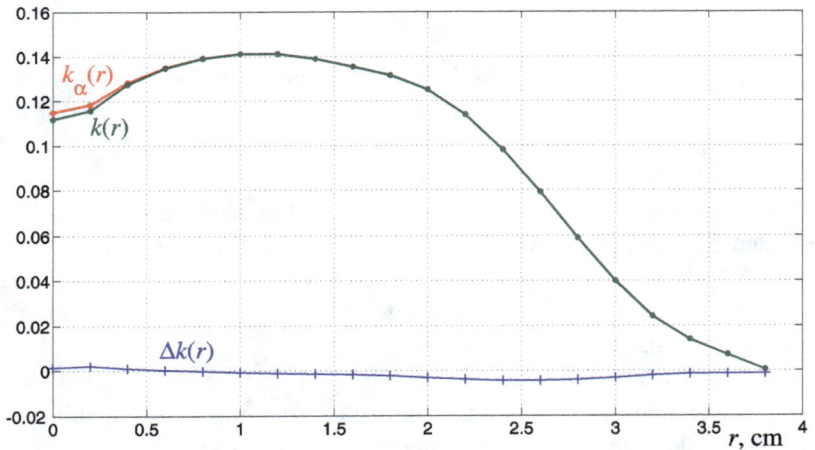

Figure 5.7.   Absorption coefficient $k(r)$ (without regularisation) and $k_\alpha(r)$ (with regularisation) after the spline smoothing of $I_m(x)$, with $n = 20$, and the errors $\Delta k(r)$ of the solution $k(r)$ cm^{-1}.

quadrature formula of left rectangles. The specificity of methods is that singular integrals are computed analytically and without peculiarities. We derived recurrence formulas for the solution, in general, on a nonuniform node mesh. The estimates of quadrature errors of the solution with regard to their sign in the absence and in the presence of errors on the right-hand side were found. In order to enhance the stability of the solution, we used Tikhonov regularisation. However, SIEs enjoy self-regularisation; therefore, it is advisable to apply the Tikhonov regularisation method only if there is a significant error ($\sim 10\%$) on the right-hand side and a rough mesh step size (i.e. when the number of nodes is small: $n \sim 10$). The method has been applied for solving infrared tomography problems.

5.4 Volterra Models for the Nonlinear Dynamics Analysis of Energy Storage Systems

Energy storage systems have been playing a key role in the power systems of the 21st century, considering the large penetrations of variable renewable energy, growth in transport electrification, and decentralisation of heating loads. Therefore, reliable real-time methods to optimise energy storage, demand response, and generation are vital for power system operations. For the convenience of the readers, this section begins with a concise review of battery energy storage and an example of battery modelling for renewable energy applications. Then, it details an adaptive approach to solving this load levelling problem with storage. A dynamic evolutionary model based on the Volterra integral equation of the first kind is used in both cases. A direct regularised numerical method is employed to find the least-cost dispatch of the battery in terms of integral equation solution. Validation against real data shows that the proposed evolutionary Volterra model effectively generalises conventional discrete integral models by taking into account both the state of health and the availability of generation/storage.

The current development of energy systems requires deep integration of renewable energy sources (RESs) and energy storage systems (ESSs) in both centralised and isolated energy systems (IRENA, 2018b). Such integration was observed for different power systems, starting with large backbone solar and wind farms, small distributed generating complexes, and small installations located directly at the customers' locality (IRENA, 2018a). The stochasticity of wind and solar generation requires storage devices acting as energy system stabilisers (Cristóbal-Monreal and Dufo-López, 2016; Tsuanyo et al., 2015; Merei et al., 2013). For large solar and wind farms, pumped stations are usually used (Bussar et al., 2014; Lunz et al., 2016). Such devices are related to medium- and long-term storage and have an installed capacity ranging from 10 to 1,000 MWh (Sauer, 2015). In practice, electrochemical batteries are widely used. They allow instant storage and generation of power to cover the load efficiently. A battery energy system connected to a centralised electrical network by means of hybrid inverters was analysed by Dufo-López (2015). Such systems allow the accumulation of energy when the tariff is low and its use during the peak hours. Thus, it is possible to reduce the annual costs for the purchase of electrical energy. Dufo-López *et al.* (2016) addressed the problem of optimising the installed capacity of batteries and solar panels in isolated power systems. Dufo-López *et al.* (2014) also compared various mathematical models predicting the service life of batteries in isolated energy systems. Ghosh *et al.* (2003) considered the photovoltaic–wind–diesel hybrid systems with hydrogen-based long-term storage, battery, and diesel generator. Hydrogen storage system was compared with the diesel-generator backup system to identify a cost-effective system. It was found that the hydrogen-based storage was especially efficient at higher latitudes depending on the seasonal renewable energy variation and fuel cost at the site of application. In the work of Dursun and Kilic (2012), the problem of optimal control of batteries in an isolated energy system with renewable energy was considered. Stevens and Corey (1996) presented the results of a process for determining battery charging efficiency. It is worth noting that distributed energy storage (DES) units are capable of enhancing the voltage stability and robustness of DC distribution networks using flexible voltage control (Li *et al.*, 2019). Li *et al.* (2018) proposed a coordinated control strategy to ensure good performance of frequency support under variations in wind conditions and the state of charge (SoC).

The main tool for analysing storage efficiency is mathematical modelling. Depending on the problem specifics, the battery models must fulfill different, sometimes contradictory, requirements. The linear models are often used to optimise the installed power of RESs and the capacity of storage devices. Bernal-Agustín *et al.* (2006) used the linear model of a battery with constant efficiency. Coordinated control loops were designed for AC/DC shipboard power systems with DES by He *et al.* (2018). The linear model of a battery presented by Lujano-Rojas *et al.* (2012) takes into account the battery inverter efficiency depending on the load factor, whereas that presented by García *et al.* (2013) takes into account its service life and the cost of its disposal. All these models are based on determining the SoC of the battery with respect to a given time interval $t \in [0, T]$. This takes the technical limitations pertaining to charge and discharge into account. For example, lead-acid (OPzV and OPzS) batteries are limited to 20–25% of their installed capacity. O. Tremblay's models implemented in

MATLAB/Simulink (Tremblay and Dessaint, 2009) allow the simulation of lead-acid, lithium-ion, nickel-cadmium, and nickel-metal hydride batteries. In addition, Tremblay's models enable us to take into account the nonlinear dependence of the no-load voltage on the SoC. A model based on the generalised Shepherd ratio was first presented by Shepherd (1965). The Shepherd model contains a nonlinear term characterising the magnitude of the polarisation voltage, depending on the current amplitude and the actual SoC of the battery.

In a real battery in idle mode, its voltage increases almost, which is generally less than the total voltage of the system. When discharge current appears, the voltage drops sharply. The presence of a nonlinear term allows us to determine the actual discharge current of the battery; however, with a numerical solution, this leads to an algebraic cycle and makes the model unstable. In the Tremblay model, the voltage value is unambiguously determined using the values of the discharge current and the actual battery charge level, which ensures sufficiently accurate simulation results for the discharge and charge modes of various types of batteries, including those used in energy systems associated with RESs. The problem of optimal power control in isolated energy systems for RESs and batteries using the Tremblay model was considered by Obukhov and Plotnikov (2017), among others. The active mass degradation of rechargeable batteries has been analysed by many authors. These and other models were developed as a methodology for categorising batteries depending on operating conditions (Svoboda *et al.*, 2007), as part of the EU project 'Benchmarking'. As a result of this project, six key indices were proposed. The numerical analysis of these indices allows us to accurately describe the processes of degradation of the active mass of batteries under the current operating conditions. Based on these indices, it is possible to determine the battery type and hardware that are the most suitable for the specific operating conditions, thereby minimising the effects of degradation processes. This technology is widely used in isolated energy systems with solar power plants.

The analysis shows that the tasks related to modelling, optimisation of installed capacity, development of new materials (Dreglea, 2012) using the Navier–Stokes models, and storage control strategies are among the key tasks at present. In most cases, in system energy studies, batteries are represented by linear discrete models. The development of new fundamental approaches describing energy storage processes is an important direction of research. Such studies should be conducted by combining robust fundamental results and practical experience.

Fredholm and Volterra (evolutionary) integral equations are at the core of many mathematical models in physics, energetics, economics, and ecology. Volterra equations are among the classic approaches in electro-analytical chemistry (Bieniasz, 2015). Brunner (1997) gives an historical overview of the results concerning the Volterra integral equations of the first kind. The theory of integral models of evolving systems was initiated in the works of L. Kantorovich, R. Solow, and V. Glushkov in the mid-20th century. This theory employs the Volterra integral equations of the first kind, where bounds of the integration interval can be functions of time; for more details, readers may refer to the monographs by Hritonenko and Yatsenko (1996), Apartsyn (2011), and Sidorov (2014b). These integral-dynamical models take into account the memory of a dynamical system, where its past impacts its future evolution, and can be employed for dynamic analysis of energy storage.

Here, we implement the energy storage dynamic analysis using the evolutionary (Volterra) dynamical models. Dynamic system analysis is carried out on a conventional isolated electric power system consisting of photovoltaic arrays and battery energy storage operated in parallel with a diesel generator backup system to serve a residential electric load.

This section is organised as follows. First, the new Volterra model of storage is presented, and a connection between the Volterra model and the conventional ampere-hour integral model is established and formulated in terms of inverse and direct problems, correspondingly. The subsequent section presents the methodology of battery modelling. The results of the model's verification against a real dataset are discussed next. This is followed by a section that deals with the application of deep learning to the dynamic analysis of a microgrid with energy storage. We conclude with final remarks, further developments, and perspectives on the proposed Volterra models of energy storage.

5.4.1   *The Volterra model of energy storage*

The Volterra equations describe the evolution of a system's state and refer to the memory expressed by an integral over a past time interval. In fact, the ampere-hour integral model (direct problem)

$$\text{SOC}(t) = \text{SOC}(0) + \int_0^t \eta(\cdot)\, i(\tau) d\tau \tag{5.61}$$

can be formulated as an inverse problem, i.e. a Volterra integral equation of the first kind, with respect to the instantaneous battery current $i(\tau)$ (which is assumed positive for charge and negative for discharge). Here, $\eta(\cdot)$ is the battery Coulombic efficiency. The SoC can be expressed either in % or in ampere-hours (or kWh). In general settings, the efficiency is a function of global time t, integration parameter τ, temperature, and other parameters. Let us now recall the generic form of Volterra integral equations:

$$\int_0^t K(t,\tau)\,x(\tau)d\tau = f(t), \quad t \in [0,T]. \tag{5.62}$$

Here, the kernel $K(t,\tau)$ and the source function $f(t)$ are assumed to be known (up to some measurement errors); $x(\tau)$ is the desired alternating power function. It is known that solutions to integral equations of the first kind can be unstable, and this represents a well-known ill-posed inverse problem. The solution of linear integral equations of the first kind is of course a classical problem and has been addressed by numerous authors. However, only a few authors have studied these equations with discontinuous kernels, $K(t,\tau)$, especially in the nonlinear case. In general, Volterra integral equations of the first kind can be solved by reducing them to equations of the second kind. Regularisation algorithms developed for Fredholm equations can also be applied, as well as direct discretisation methods, which are used here.

The theory and regularised numerical methods to relieve the ill posedness of the problem are employed in the following, as suggested by Muftahov *et al.* (2017) and Sidorov (2013c). The proposed approach to the dynamical analysis of energy storage is based on the theory of integral (hereditary) dynamic models (Sidorov, 2014b).

5.4.2 *Battery modelling*

The modelling of battery operation modes is demonstrated using the example of an isolated hybrid energy system. It is assumed that this hybrid system uses the following: photovoltaic arrays (PV), solar inverters (INV_S), battery inverters (INV_B), battery energy storage (BS), and a diesel generator (DG). Figure 5.8 shows an isolated PV system scheme used for the experiments and model validation.

Installed capacities are as follows: PV is 75 kW, INV_s is 75 kW, INV_b is 72 kW, DG is 2 × 100 kW, and BS is 384 kWh. The maximum load is

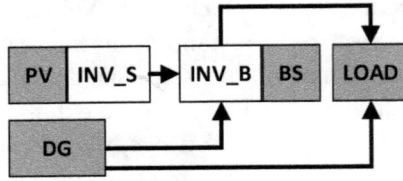

Figure 5.8.   The scheme of a microgrid with PV and battery energy storage system.

47 kW. Modelling the operation of a solar power station is performed using actinometric data recorded at the territory under consideration.

Linear model. Let us briefly describe the classic discrete mathematical model with constant efficiency commonly used in the literature. In fact, instead of the latter integral form (5.61), the following rather trivial linear discrete model is used:

$$SOC(t) = SOC(t-1) + I_s(t)\Delta t, \tag{5.63}$$

with the constraint $I_s(t) \leq r_{BS}Q_{BS}^{\max}$, where $SOC(t-1)$ (kW) is the SoC of the battery in time $t-1$, $I_s(t)$ is the alternating power function (kW), r_{BS} is the technical restriction on the charge and discharge of battery (from 20% to 40%), and $Q_{BS}^{\max}$ is the installed battery capacity (kWh). If the battery is charged, then $I_s(t)$ is multiplied by a constant efficiency of the battery and inverter. This value is assumed to be 0.8, as suggested by Stevens and Corey (1996). Δt is a discrete step (1 hour) to determine the amount of energy released to the battery. The term 'linear model' is used for the following classic model.

In order to efficiently model the storage operation, the following integral dynamical model with constraints is employed:

$$\begin{cases} \int_0^t K(t,\tau,x(\tau))\,d\tau = f(t), & 0 \leq \tau \leq t \leq T, \\[2mm] v(t) = \int_0^t x(\tau)\,d\tau, & \max_{t\in[0,T]} |v(t)| \leq v_{\max}, \\[2mm] E_{\min}(t) \leq \int_0^t v(\tau)\,d\tau \leq E_{\max}(t). \end{cases} \tag{5.64}$$

Here, the source function $f(t)$ is the energy imbalance defined as follows:

$$f(t) = f_{\mathrm{PV}}(t) - f_{\mathrm{load}}(t),$$

where $f_{PV}(t)$ is the PV generation and $f_{load}(t)$ is the electric load. This imbalance is supposed to be covered by battery storage operated in parallel with the diesel generator backup system.

In the integral model (5.64), the alternating function of changing the power $x(t)$ is the desired one. It allows for a known maximum speed of the charge v_{max}:

(1) to determine $E(t)$, which is the storage SoC under the constraints $E_{min}(t) \leq E(t) \leq E_{max}(t)$, depending on the type of storage;
(2) to determine the minimum total capacity of the storage to cover the load shortage;
(3) to calculate the number of cycles based on the behaviour of the function $E(t)$;
(4) the storage lifetime prediction for the specific region.

The problem of solving the Volterra equation in problem (5.64) is a typical inverse problem. Its integral operator is nonlinear in the general case, and the existence of a unique solution follows from Theorem 3.4.1. Moreover, it is known (see Section 1.4 in Apartsyn, 2011) that the application of discrete approximation methods to solving the Volterra equations of the first kind enjoys the self-regularisation property when the step size is matched with the amount of noise in the problem. The results of applying the regularisation procedure are discussed as follows. As done by Muftahov et al. (2017) and Sidorov (2013c), the error-resilient numerical method is employed.

The data flow diagram of developed software is shown in Figure 5.9. It consists of three main parts: the forecasting block, the block for Volterra integral equation numerical solution to find the desired SoC, and the diesel control block. Notably, for efficient application of the proposed Volterra model, it is necessary to construct accurate forecasts of the PV generation f_{PV} and the consumer electric load $f_{load}(t)$. For simplicity, these two factors are assumed to be known. In reality, they are supposed to be forecasted as time series and are therefore known up to some error level.

The load has maximum values in the winter and autumn periods, as shown in Figure 5.10.

The following conditions are met:

(1) if the alternating power function has a positive sign, then generation is sufficient for direct supply to the consumer and energy storage;

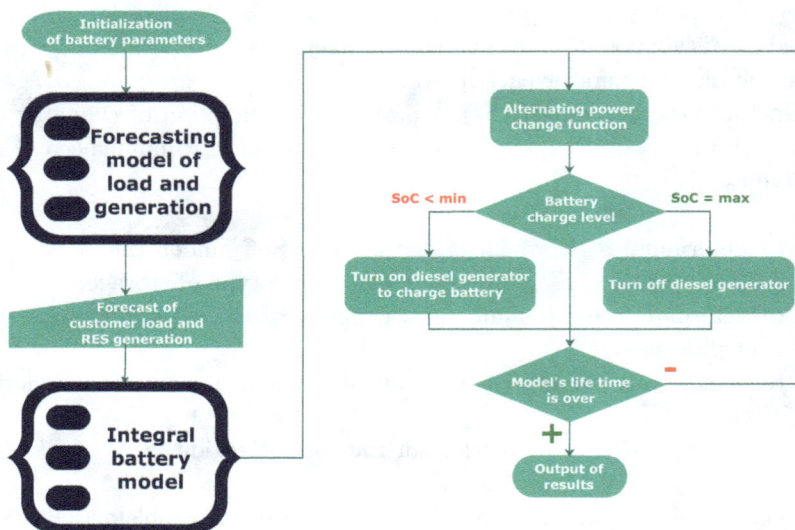

Figure 5.9. Data flow diagram of the battery storage system.

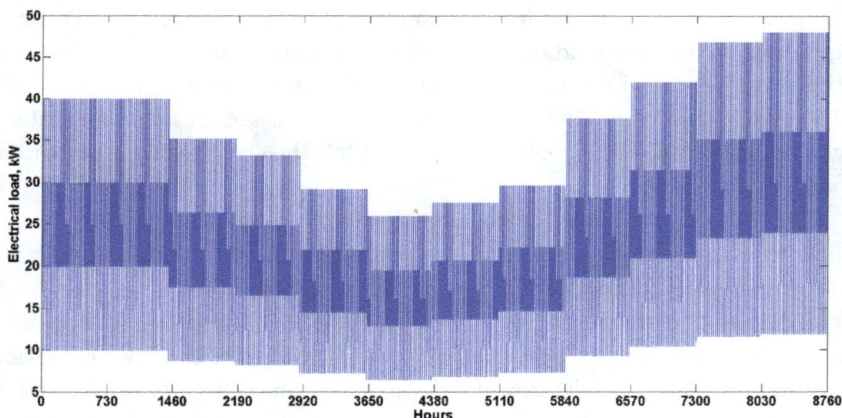

Figure 5.10. Electrical load during the year.

(2) if the alternating power function has a negative sign, then energy is not enough. Therefore, the missing energy is taken from the battery;

(3) if the battery SoC is below 20%, then the diesel generator is turned on. The generator turns on at full capacity, covering the load while charging the battery.

The system is simulated over the entire period of annual meteorological observations with a discrete step size of one hour (a total of 122,640 steps). After the simulation, the average annual value of the accumulated energy and monthly average values of SoC are calculated.

Lead-carbon batteries adapted to heavy cyclic modes were used in the simulation. The study addresses two cases:

- Case 1: the linear model of a battery with a constant efficiency.
- Case 2: the Volterra integral model with constant efficiency.

Linear model. The simulation of an isolated energy system with the linear model of a battery with a constant efficiency (5.63) has the following average hourly values of battery SoC.

Calculations demonstrated that the average annual value of the energy supplied to the battery is 48,116 kWh (PV is 20,214 kWh, and DG is 27,902 kWh). Figure 5.11 shows the average SoC of the battery for the linear model. Figure 5.12 demonstrates the SoC dynamics over a one-year period.

Volterra integral model. In the second case, the Volterra integral model with constant efficiency is used. By analogy with the first case, the simulation of the system is performed throughout the entire period of meteorological observations, followed by averaging of the necessary parameters. Let us explain the meaning of the functions in model (5.62). The right-hand side consists of the difference between the PV generation

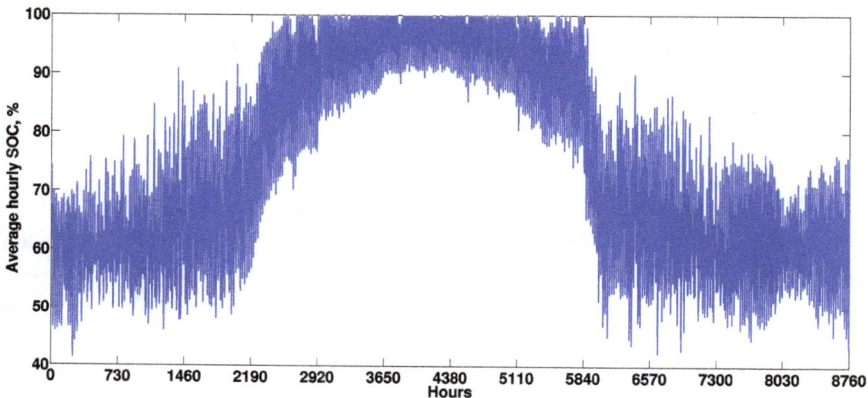

Figure 5.11. Average SoC for the linear model.

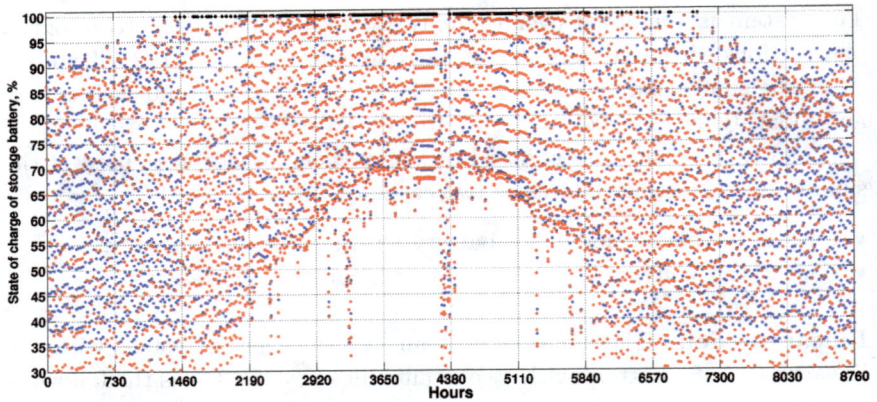

Figure 5.12.   SoC dynamics. Blue points denote charging, red points stand for discharging, and black points denote idle state at high SoC.

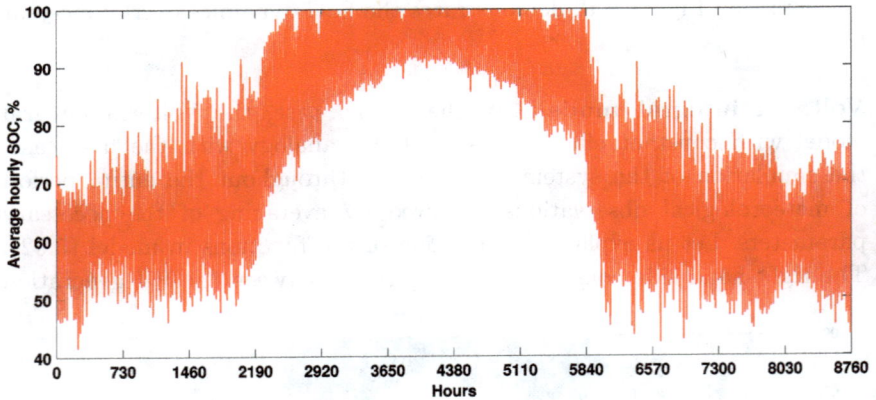

Figure 5.13.   Average SoC by the Volterra model.

and the consumer-end load. Based on the unknown alternating power function $x(t)$ and restrictions on the maximum charge/discharge rate v_{max}, the charge level of the storage is determined, which also imposes threshold limits. The efficiency of storage can depend on time and expressed by the kernel $K(t, \tau)$.

Figure 5.13 shows the average SoC of the battery for the Volterra model, which is very similar to the results shown in Figure 5.11. Both SoC curves are shown in Figure 5.14 for a 96 hour period to demonstrate the adequacy

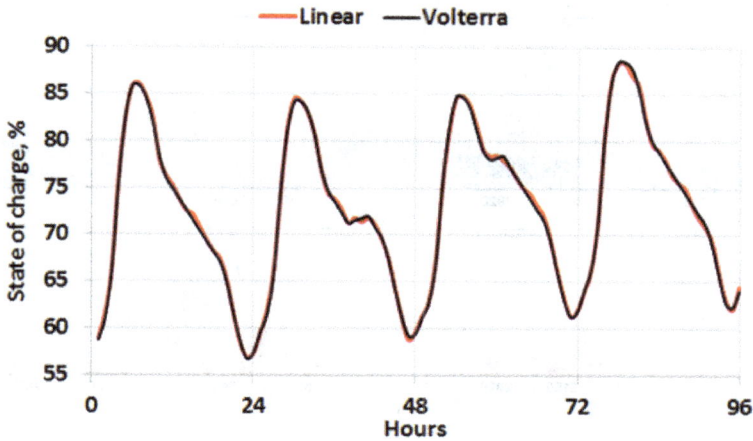

Figure 5.14.   SoC for the conventional linear model and the Volterra model.

of the employed Volterra model. The results show that the amount of energy supplied to the battery is 47,849 kWh (PV: 20185 kWh, DG: 27664 kWh). Let us now consider the detailed validation of the Volterra model, including error analysis and model robustness in input data.

5.4.3   *Model validation*

The isolated cold region energy system of Innyaly in Yakutia, Russia, was chosen for model validation. Retrospective data (over the time interval from 2005 to 2019) of the total solar radiation were used as baseline information. Solar radiation was recorded at a meteorological station located in this region. Electrical load was built according to the real data of typical days relative to each month. Let us assume that this system has two diesel generators of 100 kW each. Their real technical characteristics were used in the work. The technical data of solar panels, PV and battery inverters, rechargeable batteries are used in the following model validation. It should be noted that storage efficiency is a nonlinear function; it depends on the SoC, charging current, and temperature. A detailed account of nonlinear dependencies is a difficult task to be described by the Volterra kernel $K(t, \tau, x(\tau))$. In this study, the efficiency was assumed to be constant ($\eta = 0.8$) following Stevens and Corey (1996).

The numerical experiments were conducted for the retrospective dataset. The three following metrics were selected: $RMSE =$

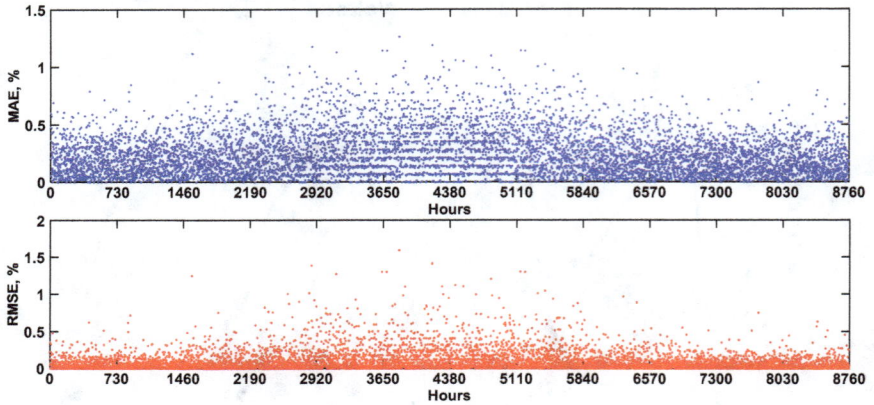

Figure 5.15.   RMSE and MAE over a one-year period.

$\sqrt{\frac{1}{n}\sum_{t=1}^{n}(x_t - \bar{x}_t)^2}$ is the root-mean-square error, the mean absolute error $MAE = \frac{1}{n}\sum_{t=1}^{n}|x_t - \bar{x}_t|$, and the average absolute error in percent $MAPE = \frac{1}{n}\sum_{t=1}^{n}\frac{|x_t - \bar{x}_t|}{\bar{x}_t}*100\%$. Here, $\bar{x}_t$ corresponds to the linear model, and x_t corresponds to the Volterra model. The *MAE, RMSE*, and *MAPE* values are 0.23%, 0.29%, and 0.31%, respectively. The results of multiple time scales are compared, and the corresponding RMSE and MAE are shown in Figure 5.15 for the one-year period. Each point corresponds to a one-hour subperiod.

The simulation of a battery using a linear model and the Volterra integral model produced similar results, as seen in Figure 5.15.

Thus, the model of the SoC of the battery based on the Volterra equation enables accurate processes description. The numerical results obtained in the calculation of the isolated energy system with PV, diesel, and the battery showed the adequacy of the Volterra model for battery modelling. On comparison, it is clear that the results are very similar to those obtained from the linear model.

An analysis of Figure 5.14 shows that the storage SoC repeats the behaviour of the alternating power function depending on the total solar radiation. It should be noted that Volterra integral equations have great potential in accounting for nonlinear processes occurring in the battery. In this case, the consideration of nonlinearity is a characteristic feature inherent to Volterra integral equations. These processes include the

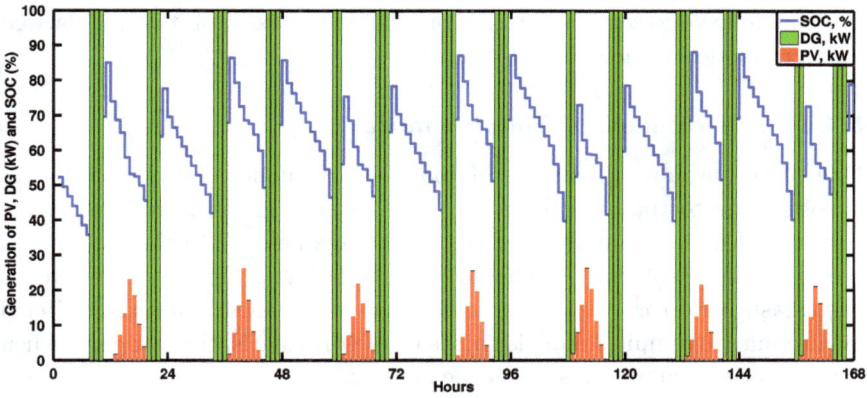

Figure 5.16.   Dynamic analysis of the SoC of the battery using the Volterra model with PV and diesel (January).

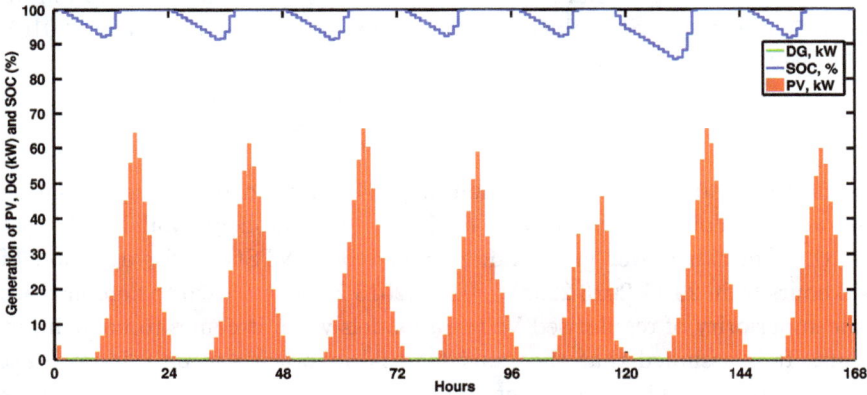

Figure 5.17.   Dynamic analysis of the SoC of the battery using the Volterra model with PV and diesel (June).

nonlinear characteristics of efficiency and cycle life depending on the depth of discharge and processes of degradation of the active mass State of Health (SoH).

As a footnote, Figures 5.16 and 5.17 show the typical dynamics of storage SoC for January and June, respectively. As can be noted, the proposed Volterra model with constraints can efficiently model the energy storage system with diesel and PV.

The following section focuses on the robustness of the regularised Volterra models to noisy input data.

5.4.4   *The regularised Volterra model*

One of the promising features of the Volterra model analysis of storage is robustness to the input data errors. As stated in the introduction, the problem of dynamic analysis of energy storage belongs to the class of ill-posed inverse problems. It should be noted that large amplifications of the measuring errors are typical of ill-posed problems. In fact, the Volterra evolutionary dynamical models enjoys the self-regularisation property when the step size is coordinated with the amount of noise in the input data (or forecast error bounds). Here, readers may refer to p. 25 in the monograph by Apartsyn (2011) for more information concerning (h, α)-regularisation.

However, the step size is not under control in this problem; therefore, the Lavrentiev α-regularisation method is the only feasible option to address this issue with solution stability. Following Muftahov *et al.* (2016), the regularised equation

$$\alpha x_\alpha(t) + \int_0^t K(t,\tau)x_\alpha(\tau)d\tau = \tilde{f}(t)$$

is used instead of the first equation in (5.64). Here, α is the regularisation parameter. Following Sidorov *et al.* (2019), the publicly available dataset of the German power grid load, provided by ENTSO-E, for a period of 48 hours from 29.11.2013 22:00 to 01.12.2013 21:00 was used to demonstrate the superiority of regularised Volterra models when it comes to noisy input data. It is assumed that the exact load is unknown and only its forecast is available. In this experiment, the storage efficiency (kernel $K(t,\tau)$) is assumed to be constant at $\eta = 0.92$.

Time series prediction is one of the emerging fields of machine and deep learning methods (Luxuan *et al.*, 2024; Liu *et al.*, 2023). The deep learning GRU algorithm (with settings: gru_1(300), gru_2(300), Dense_1(16)) was used as the base model for electric load forecasting. The Lavrentiev α-regularisation method was employed to cope with amplifications in the solution caused by inaccuracies in the forecast. Table 5.2 lists the errors analysis results using the RMSE, MAE, and MAPE metrics when the desired function is calculated based on the exact load, the GRU-based load forecast without any regularisation, and the GRU-based load forecast with

Table 5.2.   Error analysis.

	GRU reg. $\alpha = 0.422$	GRU
RMSE	96.82	754.6
MAE	75.12	537.88
MAPE	16.78%	130.05%

α-regularisation. The regularisation parameter $\alpha = 0.422$ was calculated here by the well-known discrepancy principle. Regularisation significantly reduced the errors caused by the inaccurate forecast. It should be noted that the parameter α must be dynamically adjusted to the accuracy level of a forecast.

Thus, it can be concluded that the α-regularised Volterra model is a promising tool for mathematical modelling and dynamic analysis of power grids with energy storage.

5.4.5   *The MDP-DQN model*

Optimal operation of a hybrid PV–diesel system can be formalised as a partially observable Markov decision process (MDP), where the hybrid system is considered as an agent that interacts with its environment (Duan *et al.*, 2019). The fundamental difference between the approaches is that no specific strategy is set for the model, i.e. only the dynamics of an MDP environment and the conditions of the agent's actions are considered. However, in the learning process, the agent finds the optimal policy (management strategy).

In order to approach the Markov property, the system's state $s_t \in S$ is made up of an history of the features of observations O_t^i, $i \in 1, \ldots, N_f$, where $N_f \in N$ is the total number of features. Each O_t^i is represented by a sequence of punctual observations over a chosen history of length, $h^i : O_t^i = [o_{t_{h^i+1}}^i, \ldots, o_t^i]$. At each time step, the agent observes a state variable s_t, takes an action $a_t \in A$, and moves into a state s_{t+1}. A reward signal $r_t = p(s_t, a_t, s_{t+1})$ is associated with the transition (s_t, a_t, s_{t+1}), where $p : S \times A \times S \to \mathbb{R}$ is the reward function. The γ-discounted optimal Q-value function is defined as follows:

$$Q^*(s, a) = \max_{\pi} \mathop{E}_{s_{t+1}, s_{t+2}, \ldots} \left[\sum_{k=t}^{\infty} \gamma^{k-T} r_k \,\middle|\, s_t = s, a_t = a, \pi \right].$$

We propose to approximate Q^* using a deep Q-network (DQN) because DQN models produced good results for energy microgrid management in recent works (Mocanu *et al.*, 2018; Xiao *et al.*, 2018). We adapted the approach proposed by Francois-Lavet *et al.* (2016) for a PV–hydrogen microgrid management. We used the DQN architecture, where the inputs are provided by the state vector and each separate output represents the Q-values for each discretised action. Possible actions are whether to turn on or off the DG device for covering the load and charging the battery (avoid any value of loss load whenever possible). We considered three discretised actions: (i) turn on at full capacity, (ii) keep it idle, or (iii) refill it if the DG tank is empty.

The reward function of the system corresponds to the instantaneous operational revenues r_t at time $t \in T$. The instantaneous reward signal r_t was obtained by adding the revenues generated by the hydrogen production r_DG with the penalties r^- due to the value of loss load: $r_t = r(a_t, d_t) = r^{DG}(a_t, d_t) + r^-(a_t, d_t)$, where d_t denotes the net electricity demand. From the series of rewards (r_t), we obtained the operational revenues over year y, defined as follows: $M_y = \sum_{t \in \tau_y} r_t$, where τ_y is the set of time steps belonging to year y (Francois-Lavet *et al.*, 2016). The typical behaviour of the policy for summer is illustrated in Figure 5.18. This figure demonstrates the fact that the DQN model efficiently finds the desired policy, which is quite similar to the policy (management strategy) obtained by using the Volterra model.

This section aimed to demonstrate the link between the Volterra model and widely used storage models. An integral dynamical model was proposed for the battery SoC dynamic analysis in terms of the inverse problem's efficient solution. The proposed approach has a solid theoretical basis formulated in terms of the qualitative theory of integral equations and is complemented by the development of robust numerical methods. The constructed linear Volterra model of the single battery storage with constant efficiency was verified using a real database, and its results were compared with those of the classic discrete model.

The advantages of the suggested evolutionary dynamical model are as follows:

(1) The regularised solution is robust to unavoidable errors in load and generation forecasts.
(2) It allows the definition of operating parameters of storage when various renewable energy sources and storage are used jointly.

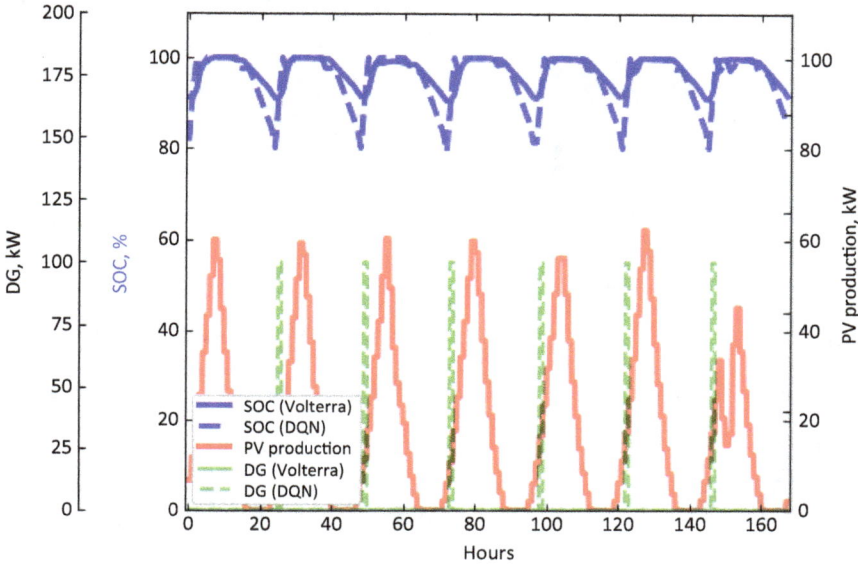

Figure 5.18. Dynamic analysis of the SoC of the battery using the Volterra and DQN models with PV and diesel (June).

(3) It considers various characteristics of storage operation, such as power, charge/discharge rate, maximum number of work cycles, and the SoC limit.

(4) The model algorithm has low computational complexity when a large number of storage units n is involved.

(5) It accounts for the nonlinear nature of storages efficiency changes as a function of SoC.

The well-developed theory of evolutionary linear and nonlinear integral equations (and their systems) with continuous and jump discontinuous kernels opens new avenues for various storage models. Unlike the classic discrete models, the proposed continuous models formulated in terms of integral equations produce error-resilient generic solutions. Volterra evolutionary models naturally take into account the temporal degradation of storage systems and their efficiency dependence on the SoC. The systems of Volterra equations can model energy storage systems that combine different energy storage technologies. The state-of-the-art machine learning methods, including DQN, combined with Volterra models will bring more insights for achieving an efficient balance of renewable sources and user demands in grids through the use of energy storage systems.

Bibliography

Abdullaev, V. and Aida-Zade, K. (2014). Numerical method of solution to loaded nonlocal boundary value problems for ordinary differential equations. *Computational Mathematics and Mathematical Physics*, 54, 1096–1109.

Acosta-Humánez, P. (2006). La teoría de Morales-Ramis y el algoritmo de Kovacic. *Lecturas Matemáticas. Volumen Especial*, 1, 21–56.

Acosta-Humánez, P. (2009). Nonautonomous Hamiltonian systems and Morales-Ramis theory I. The case $\ddot{x} = f(x,t)$. *SIAM Journal on Applied Dynamical Systems*, 8, 279–297.

Acosta-Humánez, P. (2010). *Galoisian Approach to Supersymmetric Quantum Mechanics: The Integrability Analysis of the Schrödinger Equation by Means of Differential Galois Theory*. Saarbrücken: VDM Verlag Dr. Müller.

Acosta-Humánez, P., Lázaro, J., Morales-Ruiz, J., and Pantazi, C. (2015). On the integrability of polynomial vector fields in the plane by means of Picard-Vessiot theory. *Discrete and Continuous Dynamical Systems — Series A*, 35, 1767–1800.

Acosta-Humánez, P., Morales-Ruiz, J., and Weil, J.-A. (2011). Galoisian approach to integrability of Schrodinger equation. *Report on Mathematical Physics*, 67, 305–375.

Acosta-Humánez, P. and Suazo, E. (2013). Liouvillian propagators, Riccati equation and differential Galois theory. *Journal of Physics A: Mathematical and Theoretical*, 46, 455203.

Agarwal, P., Baltaeva, U., and Alikulov, Y. (2020). Solvability of the boundary-value problem for a linear loaded integro-differential equation in an infinite three-dimensional domain. *Chaos, Solitons & Fractals*, 140, 110108.

Aidazade, K. and Abdullaev, V. (2014). On the numerical solution to loaded systems of ordinary differential equations with non-separated multipoint and integral conditions. *Numerical Analysis and Applications*, 17, 1–16.

Akhmedov, K. (1957). The analytic method of Nekrasov–Nazarov in non-linear analysis. *Uspekhi Matematicheskikh Nauk*, 12:4(76), 135–153.

Akhmetov, R., Kamensky, M., Potapov, A., Rodkina, A., and Sadovsky, B. (1982). Theory of equations of neutral type. *Proceedings of Science and Technology Mathematical Analysis*, 19, 55–126.

Albeverio, S. and Elander, N. (2002). *Operator Methods in Ordinary and Partial Differential Equations.* Operator Theory: Advances and Applications. Stockholm: Birkhäuser Verlag.

Alikhanov, A., Berezgov, A., and Shkhanukov-Lafishev, M. (2008). Boundary value problems for certain classes of loaded differential equations and solving them by finite difference methods. *Computational Mathematics and Mathematical Physics*, 48, 1581–1590.

Anderssen, R. and de Hoog, F. (1990). The theory of a general quantum system interacting with a linear dissipative system. Michael A. Golberg (ed.). In *Numerical Solution of Integral Equations.* Plenum Press, Berlin, pp. 373–410.

Apartsyn, A. S. (2011). *Nonclassical Linear Volterra Equations of the First Kind*, Vol. 39. Walter de Gruyter. Available at: doi:10.1515/9783110944976.

Baltaeva, I., Rakhimov, I., and Khasanov, M. (2022). Exact traveling wave solutions of the loaded modified Korteweg-de Vries equation. *Bulletin of Irkutsk State University. Series Mathematics*, 41, 85–95.

Baltaeva, U. (2017). The loaded parabolic-hyperbolic equation and its relation to non-local problems. *Nanosystems: Physics, Chemistry, Mathematics*, SPb, Russia, 8, 413–419.

Belotserkovskii, S. and Lifanov, I. (1993). *Method of Discrete Vortices.* Boca Raton: CRC Press, USA.

Ben Abdallah, N., Degond, P., and Mehats, F. (1987). Mathematical models of magnetic insulation. Rapport interne 20, Universite Paul Sabatier, Toulouse, France.

Ben Abdallah, N., Degond, P., and Mehats, F. (1998). Mathematical models of magnetic insulation. *Physics of Plasmas*, 5(5), 1522–1534.

Bennett, C. and Sharpley, R. (1988). Interpolation of operators. *Pure and Applied Mathematics*, Vol. 129. Boston, MA: Academic Press, Inc.

Benyahia, B., Sari, T., Cherki, B., and Harmand, J. (2012). Bifurcation and stability analysis of a two step model for monitoring anaerobic digestion processes. *Journal of Process Control*, 22(6), 1008–1019.

Bernal-Agustín, J. L., Dufo-López, R., and Rivas-Ascaso, D. M. (2006). Design of isolated hybrid systems minimizing costs and pollutant emissions. *Renewable Energy*, 31, 14, 2227–2244. Available at: doi:10.1016/j.renene. 2005.11.002.

Bernard, O., Hadj-Sadok, Z., Dochain, D., Genovesi, A., and Steyer, J. (2001). Dynamical model development and parameter identification for an anaerobic wastewater treatment process. *Biotechnology and Bioengineering*, 75(4), 424–438.

Bieniasz, L. (2015). *Modelling electroanalytical experiments by the integral equation method.* Berlin: Springer.

Boglaev, Y. P. (1979). The generalized Frobenius formula in singularly perturbed linear equations. *Soviet Mathematics — Doklady*, 20, 731–734.

Boichuk, A. A. and Samoilenko, A. M. (2016). *Generalized Inverse Operators and Fredholm Boundary-Value Problems.* Inverse and Ill-Posed Problems Series. Berlin: Walter de Gruyter GmbH.

Boikov, I. and Kudryashova, N. (2000). Approximate methods for singular integral equations in exceptional cases. *Differential Equations*, 36(9), 1360–1369. Available at: doi:10.1007/BF02754309.

Börm, S., Grasedyck, L., and Hackbusch, W. (2003). *Spectral Theory of Block Operator Matrices and Applications*. Leipzig, Germany: Max-Planck-Institut für Mathematik in den Naturwissenschaften.

Böttcher, A. and Karlovich, Y. I. (1997). *Carleson Curves, Muckenhoupt Weights, and Toeplitz Operators*. Progress in Mathematics, Vol. 154. Bassel: Birkhäuser.

Bougoffa, L., Mennouni, A., and C.Rach, R. (2013). Solving Cauchy integral equations of the first kind by the Adomian decomposition method. *Applied Mathematics and Computation*, 219(19), 4423–4433.

Brunner, H. (1997). 1896–1996: One hundred years of Volterra integral equations of the first kind. *Applied Numerical Mathematics*, 24(2), 83–93. Available at: https://doi.org/10.1016/S0168-9274(97)00013-5. *Second International Conference on the Numerical Solution of Volterra and Delay Equations*.

Bruno, A. (2000). *Power geometry in algebraic and differential equations*. Amsterdam, The Netherlands: Elsevier Science & Technology.

Bussar, C., Moos, M., Alvarez, R., Wolf, P., Thien, T., Chen, H., Cai, Z., Leuthold, M., Sauer, D. U., and Moser, A. (2014). Optimal allocation and capacity of energy storage systems in a future european power system with 100% renewable energy generation. *Energy Procedia*, 46, 40–47. Available at: doi: 10.1016/j.egypro.2014.01.156.

Castro, L. P. and Rojas, E. M. (2010). Explicit solutions of Cauchy singular integral equations with weighted Carleman shift. *Journal of Mathematical Analysis and Applications*, 371(1), 128–133.

Castro, L. P. and Rojas, E. M. (2011). On the solvability of singular integral equations with reflection on the unit circle. *Integral Equations and Operator Theory*, 70, 63–99.

Castro, L. P., Saitoh, E. M. R. S., Tuan, N. M., and Tuan, P. D. (2015). Solvability of singular integral equations with rotations and degenerate kernels in the vanishing coefficient case. *Analysis and Applications*, 13(1), 1–21. Available at: doi:10.1142/S0219530514500468.

Chadam, J. M., Peirce, A., and Yin, H. M. (1992). The blowup property of solutions to some diffusion equations with localized nonlinear reactions. *Journal of Mathematical Analysis and Applications*, 169(2), 313–328.

Chen, Y., Levine, S., and Rao, M. (2006). Variable exponent, linear growth functionals in image restoration. *SIAM Journal on Applied Mathematics*, 66(4), 1383–1406.

Chesneaux, J. (1990). *CADNA, an ADA tool for round–off error analysis and for numerical debugging*. Barcelona: Springer, pp. 1–12.

Chesneaux, J. (1992). Stochastic arithmetic properties. *IMACS Computational and Applied Mathematics*, 1, 81–91.

Chesneaux, J. and Jézéquel, F. (1998). Dynamical control of computations using the trapezoidal and Simpson's rules. *Journal of Universal Computer Science*, 4(1), 2–10.

Chuan, L. H., Mau, N. V., and Tuan, N. M. (2008). On a class of singular integral equations with the linear fractional carleman shift and the degenerate kernel. *Complex Variables and Elliptic Equations*, 53(2), 117–137.

Chuan, L. H. and Tuan, N. M. (2003). On singular integral equations with carleman shifts in the case of the vanishing coefficient. *Acta Mathematica Vietnamica*, 28(3), 319–333.

Clancey, K. and Gohberg, I. (1981). *Factorization of Matrix Functions and Singular Integral Operators*. Operator Theory: Advances and Applications, Vol. 3. Birkhäuser Verlag, Basel.

Cristóbal-Monreal, I. R. and Dufo-López, R. (2016). Optimisation of photovoltaic–diesel–battery stand-alone systems minimising system weight. *Energy Conversion and Management*, 119, 279–288. Available at: https://doi.org/10.1016/j.enconman.2016.04.050.

Cruz-Uribe, D. V. and Fiorenza, A. (2013). *Variable Lebesgue spaces, foundations and harmonic analysis*. Applied and Numerical Harmonic Analysis. Basel: Birkhäuser, Basel.

Daoud, Y. and Khidir, A. A. (2018). Modified Adomian decomposition method for solving the problem of boundary layer convective heat transfer. *Propulsion and Power Research*, 7(3), 231–237.

Daun, K., Thomson, K., Liu, F., and Smallwood, G. (2006). Deconvolution of axisymmetric flame properties using Tikhonov regularisation. *Applied Optics*, 45(19), 4638–4646. Available at: doi:10.1364/AO.45.004638.

Deimling, K. (2010). *Nonlinear functional analysis*. Berlin: Springer-Verlag.

Denisov, A. (1975). On the approximate solution of the Volterra equation of the first kind. *Journal of Computational Mathematics and Mathematical Physics*, 15(4), 1053–1056.

Deutsch, M. and Beniaminy, I. (1982). Derivative-free inversion of Abel's integral equation. *Applied Physics Letters*, 41(1), 27–28. Available at: doi:10.1063/1.93309.

Diening, L., Harjulehto, P., Hästö, P., and Ružička, M. R. (2011). *Lebesgue and Sobolev Spaces with Variable Exponents*. Lecture Notes in Mathematics, Vol. 2017. Springer, Berlin.

Diening, L. and Růžička, M. R. (2003). Calderon-Zygmund operators on generalized Lebesgue spaces $l^{p(\cdot)}$ and problems related to fluid dynamics. *Journal für die Reine und Angewandte Mathematik*, 563, 197–220.

Dikinov, K., Kerefov, A., and Nakhushev, A. (1976). A certain boundary value problem for a loaded heat equation. *Differentsialnye Uravneniya*, 12, 177–179.

Dreglea, A. and Sidorov, N. (2018). Integral equations in identification of external force and heat source density dynamics. *Buletinul Academiei de Stiinte a Republicii Moldova. Matematica*, 3(88), 68–77.

Dreglea, A. I. (2012). *Boundary Value Problems in Modeling Fiber Melt Spinning, Analytical and Numerical Methods*. Saarbrücken: Lambert Academic Publishing GmbH & Co.

Dreglea Sidorov, L., Sidorov, N., and Sidorov, D. (2023). The linear fredholm integral equations with functionals and parameters. *Buletinul Academiei de Științe a Republicii Moldova. Matematica*, 102(2), 83–91.

Duan, J., Yi, Z., Shi, D., Lin, C., Lu, X., and Wang, Z. (2019). Reinforcement-learning-based optimal control for hybrid energy storage systems in hybrid AC/DC microgrids. *IEEE Transactions on Industrial Informatics*, 1. Available at: doi:10.1109/TII.2019.2896618.

Dufo-López, R. (2015). Optimisation of size and control of grid-connected storage under real time electricity pricing conditions. *Applied Energy*, 140, 395–408. Available at: doi:10.1016/j.apenergy.2014.12.012.

Dufo-López, R., Cristóbal-Monreal, I. R., and Yusta, J. M. (2016). Optimisation of PV-wind-diesel-battery stand-alone systems to minimise cost and maximise human development index and job creation. *Renewable Energy*, 94, 280–293. Available at: doi:10.1016/j.renene.2016.03.065.

Dufo-López, R., Lujano-Rojas, J. M., and Bernal-Agustín, J. L. (2014). Comparison of different lead–acid battery lifetime prediction models for use in simulation of stand-alone photovoltaic systems. *Applied Energy*, 115, 242–253. Available at: doi:10.1016/j.apenergy.2013.11.021.

Dulov, E. and Sinitsyn, A. (2005). A numerical modelling of the limit problem for the magnetically noninsulated diode. *Applied Mathematics and Computation*, 162, 1522–1534.

Dursun, E. and Kilic, O. (2012). Comparative evaluation of different power management strategies of a stand-alone PV/wind/PEMFC hybrid power system. *International Journal of Electrical Power & Energy Systems*, 34(1), 81–89. Available at: doi:10.1016/j.ijepes.2011.08.025.

El-kalla, I. (2005). Convergence of Adomian's method applied to a class of Volterra type integro-differential equations. *International Journal of Differential Equations and Applications*, 10(2), 225–234.

Erofeev, K., Khramchenkov, E., and Biryal'tsev, E. (2019). High-performance processing of covariance matrices using GPU computations. *Lobachevskii Journal of Mathematics*, 40(5), 547–554.

Evseev, V., Clausen, S., and Fateev, A. (2013). *Optical tomography in combustion*. Unpublished: Department of Chemical Engineering. Ph.D. thesis.

Evseev, V., Fateev, A., Sizikov, V., Clausen, S., and Nielsen, K. (2011). On the development of methods and equipment for 2d-tomography in combustion. In *2011 IEEE Power and Energy Society General Meeting*, pp. 1–32.

Fedorov, A. A., Berdnikov, A. S., and Kurochkin, V. E. (2019). The polymerase chain reaction model analyzed by the homotopy perturbation method. *Journal of Mathematical Chemistry*, 57, 971–985.

Francois-Lavet, V., Gemine, Q., Ernst, D., and Fonteneau, R. (2016). Towards the minimization of the levelized energy costs of microgrids using both long-term and short-term storage devices. In *Smart Grid: Networking, Data Management, and Business Models*, CRC Press, Boca Raton, USA, pp. 295–319.

Fredholm, E. (1900). *Sur une nouvelle method pour la resolution du probleme de Direchlet*. Stockholm: Kongliga Vetenskaps-Akademiens FBRH, pp. 39–46.

Gaier, D. (1980). *Vorlesungen über Approximation im Komplexen*, Vol. 53. Birkhäuser, Basel.

Gakhov, F. D. (1937). Riemann boundary value problem. *Matematicheskii Sbornik*, 44(4), 673–683.

Gakhov, F. D. (1990). *Boundary Value Problems*. Dover Publishing, New York.

Gantmacher, F. (2005). *Applications of the Theory of Matrices*. [Trans. from Russian by J.L. Brenner]. New York: Dover Inc.

García, P., Torreglosa, J. P., Fernández, L. M., and Jurado, F. (2013). Optimal energy management system for stand-alone wind turbine/photo voltaic/hydrogen/battery hybrid system with supervisory control based on fuzzy logic. *International Journal of Hydrogen Energy*, 38(33), 14146–14158. Available at: doi:10.1016/j.ijhydene.2013.08.106. Elsevier, Amsterdam.

Ghosh, P., Emonts, B., and Stolten, D. (2003). Comparison of hydrogen storage with diesel-generator system in a PV–WEC hybrid system. *Solar Energy*, 75(3), 187–198. Available at: https://doi.org/10.1016/j.solener.2003.08.004.

Gohberg, I. and Krupnik, N. (1992a). *One-dimensional Linear Singular Integral Equations: General Theory and Applications*. Operator Theory: Advances and Applications, Vol. 54. Birkhäuser, Basel.

Gohberg, I. and Krupnik, N. (1992b). *One-dimensional Linear Singular Integral Equations: Introduction*. Operator Theory: Advances and Applications, Vol. 53. Birkhäuser, Basel.

Gokhberg, I. T. and Krein, M. (1957). Fundamental aspects of defect numbers, root numbers and indexes of linear operators. *Uspekhi Matematicheskikh Nauk*, 12:2(74), 43–118.

Golozin, G. M. (1969). *Geometric Theory of Functions of a Complex Variable*. Translations of Mathematical Monographs, Vol. 26. AMS, Rhode Island.

González-Gaxiola, O. (2019). Numerical solution for Triki-Biswas equation by Adomian decomposition method. *Optik*, 194, 163014.

Graillat, S., Jézéquel, F., Wang, S., and Zhu, Y. (2011). Stochastic arithmetic in multi precision. *Journal of Mathematics and Computer Science*, Elsevier, Amsterdam, 5, 359–375.

He, J. H. (1999). Homotopy perturbation technique. *Computer Methods in Applied Mechanics and Engineering*, Elsevier, Amsterdam, 178, 257–262.

He, J. H. (2003). Homotopy perturbation method: a new nonlinear analytical technique. *Applied Mathematics and Computation*, Elsevier, Amsterdam, 135, 73–79.

He, L., Li, Y., Shuai, Z., Guerrero, J. M., Cao, Y., Wen, M., Wang, W., and Shi, J. (2018). A flexible power control strategy for hybrid AC/DC zones of shipboard power system with distributed energy storages. *IEEE Transactions on Industrial Informatics*, 14(12), 5496–5508. Available at: doi:10.1109/tii.2018.2849201.

Hilbert, D. (1912). *Grundzüge Einer Allgemeinen Theorie der Linearen Inte Gralgleichungen*, XXVI u. 282 S. gr. 8°. von O. Blumenthal (hrsgb.) Fortschritte der mathematischen Wissenschaften in Monographien, Heft 3. Leipzig: B. G. Teubner.

Hritonenko, N. and Yatsenko, Y. (1996). *Modeling and Optimization of the Lifetime of Technologies*. Dordrecht: Kluwer Academic Publishers.

IRENA (2018a). *Off-grid Renewable Energy Solutions: Global and Regional Status and Trend*. Abu Dhabi: The International Renewable Energy Agency. Technical Report. https://www.irena.org/-/media/Files/IRENA/Agency/Publication/2018/Jul/IRENA_Off-grid_RE_Solutions_2018.

IRENA (2018b). *Renewable Energy Statistics 2018*. Abu Dhabi: The International Renewable Energy Agency. Technical Report. https://www.irena.org/publications/2018/Jul/Renewable-Energy-Statistics-2018.

Jeribi, A. (2015). *Spectral Theory and Applications of Linear Operators and Block Operator Matrices*. New YorK: Springer-Verlag.

Jézéquel, F. and Mecanique, C. (2006). A dynamical strategy for approximation methods. *Comptes Rendus Mecanique*, 334, 362–367.

Kabanikhin, S. (2012). *Inverse and Ill-posed Problems. Theory and Applications*. Berlin: de Gruyter.

Karapetiants, N. and Samko, S. (2001). *Equations with Involutive Operators*. Boston MA: Birkhäuser Boston Inc.

Karlovich, A. Y. (1998). Singular integral operators with piecewise continuous coefficients in reflexive rearragement-invariant spaces. *Integral Equations and Operator Theory*, 32, 436–481.

Karlovich, A. Y. (2009). Remark on the boundedness of the cauchy singular integral operator on variable lebesgue spaces with radial oscillating weights. *Journal of Function Spaces and Applications*, 7(3), 301–311.

Karlovich, A. Y. and Lerner, A. (2005). Commutators of singular integrals on generalized l^p spaces with variable exponent. *Publicacions Matemàtiques*, 49, 111–125.

Karlovich, A. Y. and Spitkovsky, I. M. (2014). The Cauchy singular integral operator on weighted variable Lebesgue spaces. Manuel Cepedello Boiso, Håkan Hedenmalm, Marinus A. Kaashoek, Alfonso Montes Rodríguez, Sergei Treil (eds.). In *Concrete Operators, Spectral Theory, Operators in Harmonic Analysis and Approximation*. Operator Theory: Advances and Applications, Birkhäser Publ., Basel, Vol. 236, pp. 275–291.

Kato, T. (1966). *Perturbation Theory of Linear Operators*. Berlin: Springer.

Khromov, A. (2006). Integral operators with kernels that are discontinuous on broken lines. *Matematicheskii Sbornik*, 197(11), 115–142.

Khuskivadze, G., Kokilashvili, V., and Paatasvili, V. (1998). Boundary value problems for analytic and harmonic functions in domains with nonsmooth boundary. Applications to conformal mappings. *Memoirs on Differential equations and Mathematical Physics*, 195, 14.

Kokilashvili, V. and Paatashvili, V. (2007). The Dirichlet problem for harmonic functions in the Smirnov class with variable exponent. *Georgian Mathematical Journal*, 14(2), 289–299.

Kokilashvili, V. and Paatashvili, V. (2008). The Riemann-Hilbert problem in weighted classes of Cauchy type integrals with density from $l^{p(\cdot)}(\gamma)$. *Complex Analysis and Operator Theory*, 2(4), 569–591.

Kokilashvili, V. and Paatashvili, V. (2009). The Riemann-Hilbert problem in a domain with piecewise-smooth boundaries in weighted classes of Cauchy type integrals with density from variable Lebesgue spaces. *Georgian Mathematical Journal*, 16(4), 289–299.

Kokilashvili, V. and Paatashvili, V. (2011). The Dirichlet problem for harmonic functions from variable exponent Smirnov class in domains with piecewise smooth boundary. *Journal of Mathematical Sciences*, 172(3), 1–21.

Kokilashvili, V. and Paatashvili, V. (2012). Boundary value problems for analytic and harmonic functions on nonstandard banach function spaces. *Mathematics Research Developments*. Nova Science Publishers, Hauppauge, New York.

Kokilashvili, V., Paatashvili, V., and Samko, S. (2005). Boundary value problems for analytic functions in the class of Cauchy type integrals with density in $l^{p(\cdot)}(\gamma)$. *Boundary Value Problems*, 2, 43–71.

Kokilashvili, V., Samko, N., and Samko, S. (2007). Singular operators in variable spaces $l^{p(\cdot)}(\omega, \rho)$ with oscillating weights. *Mathematische Nachrichten*, 280, 1145–1156.

Kokilashvili, V. and Samko, S. (2002). Singular integral and potentials in some Banach spaces with variable exponent. *Instituto Superior Técnico, Lisboa, Departamento de Matematicas* [Preprint], 24, 1–14.

Kokilashvili, V. and Samko, S. (2003a). Singular integral equations in the Lebesgue spaces with variable exponent. *Proceedings of A. Razmadze Mathematical Institute*, 313, 61–78.

Kokilashvili, V. and Samko, S. (2003b). Singular integral in weighted Lebesgue spaces with variable exponent. *Georgian Mathematical Journal*, 1, 145–156.

Kolmogorov, A. and Fomin, S. (1999). *Elements of the Theory of Functions and Functional Analysis*. Dover Publications, New York.

Korn, G. and Korn, T. (1961). *Mathematical Handbook for Scientists and Engineers*. McGraw-Hill Book Company, New York.

Korpusov, M. and Panin, A. (2016). *Lectures on Linear and Nonlinear Functional Analysis*, Vol. 6. Moscow: Moscow State University Press, NY, USA.

Kovacik, O. and Rakosnik, J. (1991). On spaces $l^{p(x)}$ and $w^{k,p(x)}$. *Czechoslovak Mathematical Journal*, 41(116), 592–618.

Kozlov, A., Tomin, N., Sidorov, D., Lora, E., and Kurbatsky, V. (2020). Optimal operation control of PV-biomass gasifier-diesel-hybrid systems using reinforcement learning techniques, MDPI, Basel. *Energies*, 13, 2632.

Krantz, S. (2013). *A Guide to Functional Analysis*, Vol. 49. Washington: MAA Press.

Krasnosel'skii, M. (1964). *Topological Methods in the Theory of Nonlinear Integral Equations*. Oxford: Pergamon Press.

Kravchenko, V. G. and Litvinchuk, G. S. (1994). *Introduction to the Theory of Singular Integral Operators with Shift*. Amsterdam: Kluwer Academic Publishers.

Krylov, V. (2005). *Approximate Calculation of Integrals*. Dover Books on Mathematics. Dover Publications, Mineola, NY. Available at: doi:10.1137/1. 9780898718836.

Krylov, V., Bobkov, V., and Monastyrnyi, P. (1977). *Computational Methods* (in Russian), Vol. 2. Nauka, Moscow, USSR.

Lample, B. and Rosenwasser, E. (2010). Using the Fredholm resolvent for computing the h_2-norm of linear periodic systems. *International Journal of Control*, 83(96), 1868–1884.

Langmuir, I. and Compton, K. (1931). Electrical discharges in gases part II. Fundamental phenomena in electrical discharges. *Reviews of Modern Physics*, 3, 191–257. https://api.semanticscholar.org/CorpusID:123500054.

Lavrent'ev, M. and Savel'ev, L. (1995). *Linear Operators and Ill-Posed Problems*. Springer, Berlin.

Lavrent'ev, M. and Savel'ev, L. (2006). *Operator Theory and Ill-Posed Problems*. Inverse and Ill-Posed Problems Series, Vol. 50. Walter de Gruyter GmbH, Berlin, Germany.

Leontyev, R. (2013). *Nonlinear Equations in Banach Spaces with a Vector Parameter in Irregular Cases*. Irkutsk, Russia: Irkutsl State University Publishing.

Li, Y., He, L., Liu, F., Li, C., Cao, Y., and Shahidehpour, M. (2019). Flexible voltage control strategy considering distributed energy storages for dc distribution network. *IEEE Transactions on Smart Grid*, 10(1), 163–172. Available at: doi:10.1109/TSG.2017.2734166.

Li, Y., He, L., Liu, F., Tan, Y., Cao, Y., Luo, L., and Shahidehpour, M. (2018). A dynamic coordinated control strategy of WTG-ES combined system for short-term frequency support. *Renewable Energy*, 119, 1–11. Available at: https://doi.org/10.1016/j.renene.2017.11.064, http://www.sciencedirect.com/science/article/pii/S0960148117311655.

Litvinchuk, G. (2000). Solvability theory of boundary value problems and singular integral equations with shift. *Mathematics and Its Applications*, Vol. 523. Kluwer, Amsterdam.

Litvinchuk, G. S. and Spitkovsky, I. M. (1987). Factorization of measurable matrix functions. *Operator Theory: Advances and Applications*, Vol. 25. Birkhäuser Verlag, Basel.

Liu, F., Liu, Q., Tao, Q., Huang, Y., Li, D., and Sidorov, D. (2023). Deep reinforcement learning based energy storage management strategy considering prediction intervals of wind power. *International Journal of Electrical Power and Energy Systems*, Elsevier, 145, 108608. Available at: https://doi.org/10.1016/j.ijepes.2022.108608.

Loginov, B. and Sidorov, N. (1992). Group symmetry of the Lyapunov–Schmidt branching equation and iterative methods in the bifurcation point problem. *Mathematics of the USSR-Sbornik*, American Mathematical Society (AMS), Providence, RI, 73(1), 67–77.

Lujano-Rojas, J. M., Monteiro, C., Dufo-López, R., and Bernal-Agustín, J. L. (2012). Optimum load management strategy for wind/diesel/battery hybrid power systems. *Renewable Energy*, 44, 288–295. Available at: doi:10.1016/j.renene.2012.01.097.

Lunz, B., Stocker, P., Eckstein, S., Nebel, A., Samadi, S., Erlach, B., Fischedick, M., Elsner, P., and Sauer, D. U. (2016). Scenario-based comparative assessment of potential future electricity systems — a new methodological approach using Germany in 2050 as an example. *Applied Energy*, 171, 555–580. Available at: doi:10.1016/j.apenergy.2016.03.087.

Lusternik, L. (1956). Some issues of nonlinear functional analysis. *Russian Mathematical Surveys*, 6(11), 145–168.

Luxuan, Y., Ting, G., Wei, W., Min, D., Cheng, F., and Jinqiao, D. (2024). Multi-task meta label correction for time series prediction. *Pattern Recognition*, 150, 110319. Available at: https://doi.org/10.1016/j.patcog.2024.110319.

Machado Higuera, M. (2015). *Existencia de super y sub soluciones, estabilidad y bifurcación para un modelo matemático de digestión anaerobia para la producción de biogás*. Unpublished: Xalapa: Universidad Veracruzana de México. Ph.D. thesis.

Machado Higuera, M. and Sinitsyn, A. V. (2015). Existence of lower and upper solutions in reverse order with respect to a variable in a model of acidogenesis to anaerobic digestion. *Bulletin of the South Ural State University. Series Mathematical Modelling, Programming & Computer Software*, 8(2), 55–68.

Mahmoudi, Y. (2014). A new modified Adomian decomposition method for solving a class of hypersingular integral equations of second kind. *Journal of Computational and Applied Mathematics*, 255, 737–742.

Malozëmov, V., Monaco, M., and Petrov, A. (2002). The Frobenius formula and Sherman–Morrison formulas and related matters. *Computational Mathematics and Mathematical Physics*, 42(10), 1403–1409.

Mariet, F., Bernard, O., Ras, M., Lardon, L., and Steyer, J. (2011). Modeling anaerobic digestion of microalgae using ADM1. *Bioresource Technology*, 102(13), 6823–6829.

Merei, G., Berger, C., and Sauer, D. U. (2013). Optimization of an off-grid hybrid PV–wind–diesel system with different battery technologies using genetic algorithm. *Solar Energy*, 97, 460–473. Available at: doi:10.1016/j.solener.2013.08.016.

Mikhlin, S. G. and Prössdorf, S. (1986). *Singular Integral Operators*. Berlin: Springer-Verlag, 528 p.

Minerbo, G. and Levy, M. (1969). Inversion on abel's integral equation by means of orthogonal polynomials. *SIAM Journal on Numerical Analysis*, 6(4), 598–616.

Mocanu, E., Mocanu, D. C., Nguyen, P. H., Liotta, A., Webber, M. E., Gibescu, M., and Slootweg, J. G. (2018). On-line building energy optimization using deep reinforcement learning. *IEEE Transactions on Smart Grid*, 1. Available at: doi:10.1109/TSG.2018.2834219.

Morozov, V. (1984). *Methods for Solving Incorrectly Posed Problems*. Springer, Berlin.

Muftahov, I., Tynda, A., and Sidorov, D. (2017). Numeric solution of Volterra integral equations of the first kind with discontinuous kernels. *Journal of Computational and Applied Mathematics*, 313, 119–128. Available at: doi: 10.1016/j.cam.2016.09.003.

Muftahov, I. R., Sidorov, D. N., and Sidorov, N. A. (2016). Lavrentiev regularization of integral equations of the first kind in the space of continuous functions. *The Bulletin of Irkutsk State University. Series Mathematics*, 15, 62–77.

Muskhelishvili, N. I. (1968). *Singular Integral Equations*, Vol. 154. Nauka, Moscow, USSR.

Muskhelishvili, N. I. (1941). Applications of integrals of Cauchy type to a class of singular integral equations. *Transactions of Mathematical Institute Tbilissi*, 10, 1–43.

Nahushev, A. (2012). *Loaded Equations and Their Applications*. Moscow: Nauka.

Nekrasov, A. (1951). *The Exact Theory of Steady Waves on the Surface of a Heavy Fluid* [Transaction from Russian Izdatel'stvo Akademii Nauk SSSR]. University of Wisconsin. MRC Report iv0813.

Noeiaghdam, S., Dreglea, A., He, J., Avazzadeh, Z., Suleman, M., Araghi, M. A. F., Sidorov, D., and Sidorov, N. (2020a). Error estimation of the homotopy perturbation method to solve second kind Volterra integral equations with piecewise smooth kernels: application of the CADNA library. *Symmetry*, 12, 1730.

Noeiaghdam, S., Dreglea, A., He, J. H., Avazzadeh, Z., Suleman, M., Fariborzi Araghi, M. A., Sidorov, D., and Sidorov, N. (2020b). Error estimation of the homotopy perturbation method to solve second kind Volterra integral equations with piecewise smooth kernels: application of the CADNA library. *Symmetry*, 12, 1730.

Noeiaghdam, S., Sidorov, D., Wazwaz, A.-M., Sidorov, N., and Sizikov, V. (2021). The numerical validation of the Adomian decomposition method for solving Volterra integral equation with discontinuous kernels using the CESTAC method. *Mathematics*, 9(3), 1–15.

Novin, R., Fariborzi Araghi, M. A., and Mahmoudi, Y. (2018). A novel fast modification of the Adomian decomposition method to solve integral equations of the first kind with hypersingular kernels. *Journal of Computational and Applied Mathematics*, 343(1), 619–634.

Obukhov, S. and Plotnikov, I. (2017). Simulation model of operation of autonomuus photovoltaic plant under actual operation conditions. *Geo Assets Engineerings*, 328(6), 38–51.

Paatashvili, V. (2010). Haseman's problems in classes of functions representable by the Cauchy type integral with density from $l^{p(\cdot)}(\gamma; \rho)$. *Memoirs on Differential Equations and Mathematical Physics*, 50, 139–162.

Plemelj, J. (1908). Ein ergänzungssatz zur Cauchyschen integraldarstellung analytischer funktionen, randwerte betreffend. *Monatshefte für Mathematik und Physik*, 19, 205–210.

Poincaré, H. (1910). *Leçons de mécanique céleste professées a la Sorbonne. Tome III: Théorie des marées, rédigee par R. Fichot.* IV 4 + 469 S. 8^0. Paris: Gauthier-Villars.

Preobrazhensky, N. and Pikalov, V. (1982). *Unstable Problems of Plasma Diagnostics*. Nauka, Moscow, Soviet Union.

Rafeiro, H. and Rojas, E. (2014). *Espacios de Leguesge con Exponente Variable. Un espacio de Banach de funciones medibles.* Caracas: Editorial IVIC.

Riemann, B. (1867). *Grundlagen Für Eine Allgemeine Theorie der Funktionen Einer Veränderlichen Komplexen Grösse.* Leipzig: Worke.

Rojas, E. (2015). Boundary value problems and singular integral equations on Banach function spaces. *Dissertationes Mathematicae,* DOI: 10.4064/dm742-12-2015, IM PAN, Warszawa, Poland, 28(512), 1–42.

Saadamandi, A. and Dehghan, M. (2008). A collocation method for solving abel's integral equations of first and second kinds. *Zeitschrift für Naturforschung,* 63, 752–756. Available at: doi:10.1515/zna-2008-1202.

Saelao, J. and Yokchoo, N. (2020). The solution of Klein-Gordon equation by using modified Adomian decomposition method. *Mathematics and Computers in Simulation,* 171, 94–102.

Salinas, E., Muñoz, R., Sosa, J., and López, B. (2013). Analysis to the solutions of Abel's differential equations of the first kind under transformation $y = u(x)z(x) + v(x)$. *Applied Mathematical Sciences,* 7, 2075–2092.

Sauer, D. U. (2015). Chapter 2 — classification of storage systems. In Moseley, P. T. and Garche, J. (eds.), *Electrochemical Energy Storage for Renewable Sources and Grid Balancing.* Amsterdam: Elsevier, pp. 13–21.

Shepherd, C. (1965). Design of primary and secondary cells. p. 2. an equation describing battery discharge. *Journal of Electrochemical Society,* 2, 657–664.

Sialve, B., Bernet, N., and Bernard, O. (2009). Anaerobic digestion of microalgae as a necessary step to make microalgal biodiesel sustainable. *Biotechnology Advances,* 27(4), 409–416.

Sidorov, D. (2013a). *Methods of Analysis of Integral Dynamic Models: Theory and Applications.* Irkutsk: Irkutsk State University Publishing.

Sidorov, D. (2013b). Solvability of systems of Volterra integral equations of the first kind with piecewise continuous kernels. *Russian Mathematics,* 57, 54–63.

Sidorov, D. (2013c). On parametric families of solutions of Volterra integral equations of the first kind with piecewise smooth kernel. *Differential Equations,* 49(2), 210–216. Available at: doi:10.1134/s0012266113020079.

Sidorov, D. (2014a). Existence and blow-up of Kantorovich principal continuous solutions of nonlinear integral equations. *Differential Equations,* 50, 1217–1224.

Sidorov, D. (2014b). *Integral Dynamical Models: Singularities, Signals and Control.* World Scientific Nonlinear Science Series A. Singapore: World Scientific Publishing.

Sidorov, D., Muftahov, I., Tomin, N., Karamov, D., Panasetsky, D., Dreglea, A., Liu, F., and Foley, A. (2020a). A dynamic analysis of energy storage with renewable and diesel generation using volterra equations. *IEEE Transactions on Industrial Informatics,* 16(5), 3451–3459.

Sidorov, D. and Sidorov, N. (2012). Convex majorants method in the theory of nonlinear Volterra equations. *Banach Journal of Mathematical Analysis,* 16(1), 1–10.

Sidorov, D. and Sidorov, N. (2017). Solution of irregular systems of partial differential equations using skeleton decomposition of linear operators. *Vestnik Yuzhno-Ural'skogo Universiteta. Seriya Matematicheskoe Modelirovanie i Programmirovanie*, 10(2), 63–73.

Sidorov, D., Tao, Q., Muftahov, I., Zhukov, A., Karamov, D., Dreglea, A., and Liu, F. (2019). Energy balancing using charge/discharge storages control and load forecasts in a renewable-energy-based grids. In *2019 38th Chinese Control Conference (CCC)*, Vol. 1, pp. 1–6. Available at: doi:10.23919/ChiCC.2019.8483557.

Sidorov, D., Tynda, A., and Muftahov, I. (2014). Numerical solution of Volterra integral equations of the i kind with piecewise continuous kernels. *Vestnik South Ural State University Series Mathematical Modelling Programming and Computer Software*, 7(3), 107–115.

Sidorov, N. (1984). A class of degenerate differential equations with convergence. *Mathematical Notes of the Academy of Sciences of the USSR*, 35, 300–305.

Sidorov, N. (1995). Explicit and implicit parametrizations in the construction of branching solutions by iterative methods. *Sbornik: Mathematics*, 186(2), 297–310.

Sidorov, N. (2001). Parametrization of simple branching solutions of full rank and iterations in nonlinear analysis. *Russian Mathematics (Iz. VUZ)*, 45(9), 55–61.

Sidorov, N. (2022). Special issue editorial "solvability of nonlinear equations with parameters: branching, regularization, group symmetry and solutions blow-up". *Symmetry*, 14, 226.

Sidorov, N., Dreglea, A., and Sidorov, D. (2021). Generalisation of the Frobenius formula in the theory of block operators on normed spaces. *Mathematics*, 9, 23.

Sidorov, N. and Dreglea Sidorov, L. (2022). Integral equations. In *On Bifurcation Points of the Solution of the Hammerstein Integral Equation with Loads*. Irkutsk: ISU Publishing, pp. 41–44.

Sidorov, N. and Dreglea Sidorov, L. (2023). On the solution of Hammerstein integral equations with loads and bifurcation parameters. *Bulletin of the Irkutsk State University. Series Mathematics*, 43, 78–90.

Sidorov, N., Loginov, B., Sinitsyn, A., and Falaleev, M. (2002). *Lyapunov-Schmidt Methods in Nonlinear Analysis and Applications*. Cham: Springer Publishing.

Sidorov, N. and Sidorov, D. (2014). On the solvability of one class of operator operators Volterra equations of the first kind with piecewise continuous kernel. *Mathematical Notes*, 96(5–6), 811–826.

Sidorov, N. and Sidorov, D. (2021). Nonlinear Volterra equations with loads and bifurcation parameters: existence theorems and solution construction. *Differential Equations*, 57, 1654–1664.

Sidorov, N. and Sidorov, D. (2022). Branching solutions of the Cauchy problem for nonlinear loaded differential equations with bifurcation parameters. *Mathematics*, 10(12), 2134.

Sidorov, N., Sidorov, D., and Dreglea, A. (2020b). Solvability and bifurcation of solutions of nonlinear equations with Fredholm operator. *Symmetry*, 12, 912.

Sidorov, N., Sidorov, D., and Krasnik, A. (2010). Solution of Volterra operator-integral equations in the nonregular case by the successive approximation method. *Differential Equations*, 46, 882–889.

Sidorov, N., Sidorov, D., and Muftahov, I. (2015). Perturbation theory and the Banach–Steinhaus theorem for regularisation of the linear equations of the first kind. *Bulletin of Irkutsk State University. Series Mathematics*, 14, 82–99.

Sidorov, N., Sidorov, D., and Sinitsyn, A. (2020c). *Toward General Theory of Differential-Operator and Kinetic Models*. World Scientific Series on Nonlinear Science Series A, Vol. 97. Singapore: World Scientific Publishing.

Sidorov, N. and Trenogin, V. (1976). A certain approach to the problem of regularisation on the basis of the perturbation of linear operators. *Mathematical Notes of the Academy of Sciences of the USSR*, 20, 976–979.

Sidorov, N. and Trenogin, V. (1981). Regularization of linear equations by the application of perturbation theory. *Differential Equations*, 16, 1303–1311.

Sidorov, N. and Trenogin, V. (2003). *Bifurcation Points of Nonlinear Equations*. Moscow: FIZMATLIT, pp. 255–285.

Sidorov, N. and Trufanov, A. (2009). Nonlinear operator equations with a functional perturbation of the argument of neutral type. *Differential Equations*, 45, 1840–1844.

Simonenko, I. B. (1964). The Riemann boundary value problem for n pairs functions with measurable coefficients and its application to the investigation of singular integral operators in the spaces l^p with weight. *Izvestiya AN SSSR Seriya Matematicheskaya*, 28(2), 277–306.

Simonenko, I. B. (1968). Some general questions in the theory of the Riemann boundary value problem. *Mathematics of the USSR-Izvestiya*, 2, 1091–1099.

Singh, V., Pandey, R., and Singh, O. (2009). New stable numerical solutions of singular integral equations of abel type by using normalized Bernstein polynomials. *Applied Mathematical Sciences*, 3, 241–255. Available at: doi: 10.1016/j.jqsrt.2009.07.007.

Sizikov, V. and Sidorov, D. (2016). Generalized quadrature for solving singular integral equations of abel type in application to infrared tomography. *Applied Numerical Mathematics*, 106, 69–78. Available at: https://doi.org/ 10.1016/j.apnum.2016.03.004.

Sizikov, V., Smirnov, A., and Fedorov, B. (2004). Numerical solution of the singular abel integral equation by the generalized quadrature method. *Izvestiya Vysshikh Uchebnykh Zavedenii. Matematika*, 48(8), 59–66.

Stevens, J. W. and Corey, G. P. (1996). A study of lead-acid battery efficiency near top-of-charge and the impact on PV system design. In *Conference Record of the Twenty Fifth IEEE Photovoltaic Specialists Conference — 1996*, pp. 1485–1488. Available at: doi:10.1109/PVSC.1996.564417.

Svoboda, V., Wenzl, H., Kaiser, R., Jossen, A., Baring-Gould, I., Manwell, J., Lundsager, P., Bindner, H., Cronin, T., Nørgård, P., Ruddell, A., Perujo, A., Douglas, K., Rodrigues, C., Joyce, A., Tselepis, S., van der Borg, N., Nieuwenhout, F., Wilmot, N., Mattera, F., and Sauer, D. U. (2007). Operating conditions of batteries in off-grid renewable energy systems. *Solar Energy*, 81(11), 1409–1425. Available at: doi:10.1016/j.solener.2006.12.009.

Tikhonov, A. and Arsenin, V. (1995). *Solutions of Ill-Posed Problems*. Winston and Sons, New York.

Tikhonov, A., Goncharsky, A., Stepanov, V., and Yagola, A. (1995). *Numerical Methods for the Solution of Ill-Posed Problems*. Springer, Berlin.

Tikhonov, A., Leonov, A., and Yagola, A. (1997). *Nonlinear Ill-Posed Problems*. Chapman and Hall/CRC, Boca Baton, Florida.

Tomin, N., Shakirov, V., Kozlov, A., Sidorov, D., Kurbatsky, V., Rehtanz, C., and Lora, E. E. (2022). Design and optimal energy management of community microgrids with flexible renewable energy sources. *Renewable Energy*, 183, 903–921. Available at: https://doi.org/10.1016/j.renene.2021.11.024.

Tremblay, O. and Dessaint, L.-A. (2009). Experimental validation of a battery dynamic model for ev applications. *World Electric Vehicle Journal*, 3(2), 289–298.

Trenogin, V. (1980). *Functional Analysis*. Moscow: Nauka.

Trenogin, V. (1996). Locally invertible operators and the method of continuation with respect to parameter. *Functional Analysis and Its Applications*, 30(2), 147–148.

Trénoguine, V. (1985). *Analyse fonctionnelle [traduit du russe par V. Kotliar]*. Moscow, USSR: MIR Publishing.

Tretter, C. (2008). *Spectral Theory of Block Operator Matrices and Applications*. London: Imperial College Press.

Tricomi, F. G. (1985). *Integral Equations*. Dover Publishing, New York.

Tsuanyo, D., Azoumah, Y., Aussel, D., and Neveu, P. (2015). Modeling and optimization of batteryless hybrid PV (photovoltaic)/diesel systems for off-grid applications. *Energy*, 86, 152–163. Available at: doi:10.1016/j.energy.2015.03.128.

Tuan, N. M. (1996). On a class of singular integral equations with rotation. *Acta Mathematica Vietnamica*, 21(2), 201–211.

Vainberg, M. and Trenogin, V. (1964). *Theory of Branching of Solutions of Nonlinear Equations*. Leyden: Wolters-Noordhoff B.V.

Vapnik, V. (2006). *Estimation of Dependences Based on Empirical Data*. New York: Springer-Verlag.

Vekua, N. P., Gibbs, A. G., and Ferziger, J. H. (1967). *Systems of Singular Integral Equations*. Holland: P. Noordhoff Groningen.

Verlan', A. and Sizikov, V. (1986). *Integral Equations: Methods, Algorithms, Programs*. Kiev: Naukova Dumka.

Vignes, J. (1993). A stochastic arithmetic for reliable scientific computation. *Mathematics and Computers in Simulation*, 35, 233–261.

Vignes, J. (2004). Discrete stochastic arithmetic for validating results of numerical software. *Special Issue of Numerical Algorithms*, 37, 377–390.

Voskoboynikov, Y., Preobrazhensky, N., and Sedel'nikov, A. (1984). *Mathematical Treatment of Experiment in Molecular Gas Dynamics.* Moscow: Nauka.

Wazwaz, A.-M. (2009). *Partial Differential Equations and Solitary Waves Theory.* Berlin: Springer–Verlag.

Wazwaz, A.-M. (2011). *Linear and Nonlinear Integral Equations, Methods and Applications.* Berlin: Springer–Verlag.

Wazwaz, A.-M., Rach, R., and Duan, J.-S. (2013). Adomian decomposition method for solving the Volterra integral form of the Lane–Emden equations with initial values and boundary conditions. *Journal of Computational and Applied Mathematics*, 219(10), 5004–5019.

Xiao, L., Xiao, X., Dai, C., Peng, M., Wang, L., and Poor, H. V. (2018). Reinforcement learning-based energy trading for microgrids. *CoRR* abs/1801.06285, arXiv:1801.06285, http://arxiv.org/abs/1801.06285.

Zhuravlev, V., Fomin, N., and Zabrodskiy, P. (2019). Conditions of solvability and representation of the solutions of equations with operator matrices. *Ukrainian Mathematical Journal*, 71, 537–553.

Index

Series on Advances in Mathematics for Applied Sciences

Aims and Scope

This series reports on new developments in mathematical research related to new analytical and numerical methods, mathematical modeling in the applied physical and natural sciences as well as in economics, and quantitative and qualitative analysis approaches for the mathematical models. Topics covered include modelling, constitutive theories, fluid and solid mechanics, kinetic and transport theories. The series ranges from monographs to lecture notes, quality conference proceedings and collections of papers. The high quality of research, the novelty of mathematical tools, and the potential for frontier problems will be the guidelines for the selection of the content for this series.

Instructions for Authors

Submission of proposals should be addressed to the editors-in-charge or to any member of the editorial board. In the latter, the authors should also notify the proposal to one of the editors-in-charge. Acceptance of books and lecture notes will generally be based on the description of the general content and scope of the book or lecture notes as well as on sample of the parts judged to be more significantly by the authors.

Acceptance of proceedings will be based on relevance of the topics and of the lecturers contributing to the volume.

Acceptance of monograph collections will be based on relevance of the subject and of the authors contributing to the volume.

Authors are urged, in order to avoid re-typing, not to begin the final preparation of the text until they received the publisher's guidelines. They will receive from World Scientific the instructions for preparing camera-ready manuscript.

Series on Advances in Mathematics for Applied Sciences

ISSN: 1793-0901

Published*:

*To view the complete list of the published volumes in the series, please visit:
https://www.worldscientific.com/series/samas